MIX
Papier aus verantwortungsvollen Quellen
Paper from responsible sources
FSC® C105338

Blessed Olalekan Oyebola

Design of an Audio Multitone Refiner, Simulation of Audio Frequencies & Analysis Using Active Filter

Anchor Academic
Publishing

Oyebola, Blessed Olalekan: Design of an Audio Multitone Refiner, Simulation of Audio Frequencies & Analysis Using Active Filter, Hamburg, Anchor Academic Publishing 2017

Buch-ISBN: 978-3-96067-135-0
PDF-eBook-ISBN: 978-3-96067-635-5
Druck/Herstellung: Anchor Academic Publishing, Hamburg, 2017

Bibliografische Information der Deutschen Nationalbibliothek:
Die Deutsche Nationalbibliothek verzeichnet diese Publikation in der Deutschen Nationalbibliografie; detaillierte bibliografische Daten sind im Internet über http://dnb.d-nb.de abrufbar.

Bibliographical Information of the German National Library:
The German National Library lists this publication in the German National Bibliography. Detailed bibliographic data can be found at: http://dnb.d-nb.de

All rights reserved. This publication may not be reproduced, stored in a retrieval system or transmitted, in any form or by any means, electronic, mechanical, photocopying, recording or otherwise, without the prior permission of the publishers.

Das Werk einschließlich aller seiner Teile ist urheberrechtlich geschützt. Jede Verwertung außerhalb der Grenzen des Urheberrechtsgesetzes ist ohne Zustimmung des Verlages unzulässig und strafbar. Dies gilt insbesondere für Vervielfältigungen, Übersetzungen, Mikroverfilmungen und die Einspeicherung und Bearbeitung in elektronischen Systemen.

Die Wiedergabe von Gebrauchsnamen, Handelsnamen, Warenbezeichnungen usw. in diesem Werk berechtigt auch ohne besondere Kennzeichnung nicht zu der Annahme, dass solche Namen im Sinne der Warenzeichen- und Markenschutz-Gesetzgebung als frei zu betrachten wären und daher von jedermann benutzt werden dürften.

Die Informationen in diesem Werk wurden mit Sorgfalt erarbeitet. Dennoch können Fehler nicht vollständig ausgeschlossen werden und die Diplomica Verlag GmbH, die Autoren oder Übersetzer übernehmen keine juristische Verantwortung oder irgendeine Haftung für evtl. verbliebene fehlerhafte Angaben und deren Folgen.

Alle Rechte vorbehalten

© Anchor Academic Publishing, Imprint der Diplomica Verlag GmbH
Hermannstal 119k, 22119 Hamburg
http://www.diplomica-verlag.de, Hamburg 2017
Printed in Germany

CONTENTS

ABSTRACT .... *iii*

FOREWORD .... *iv*

CHAPTER ONE

INTRODUCTION 1

    1.1   What Is Filter? 1

    1.2   Fundamentals of Filters (Low-Pass) 2

    1.3   Superiority of Active Filters over Passive Filters 3

    1.4   Basic Op-Amp and Feedback Theory 4

    1.5   Overview: Block Diagram 5

    1.6   Terms Definition 5

CHAPTER TWO

LITERATURE REVIEW 6

    2.1   Transfer Functions 8

    2.2   Butterworth Low Pass Filters 11

    2.3   Tschebyscheff Low Pass Filters 11

    2.4   Bessel Low Pass Filters 11

    2.5   Quality Factor Q 12

    2.6   Précis 12

    2.7   Low Pass Filter Design 12

    2.8   First Order Low Pass Filter 13

    2.9   Sallen-key Topology 15

    2.10   Multiple Feedback Topologies 17

    2.11   Higher Order Low-pass Filters 17

    2.11.1   Second Order Filter 17

    2.11.2   Third Order Filter 18

    2.12   High Pass Filter Design 18

    2.13   Multiple Feedback Topologies 19

2.14	Band-Pass Filter Design	19
2.14.1	Second Order Band Pass Filter	20
2.14.2	Sallen Key Topology	20
2.14.3	Multiple Feedback Topology	21
2.15	Tone Control (Baxandall) Circuit	25

CHAPTER THREE

DESIGN 27

3.1	Filter Architecture	28
3.2	Op-Amp Selections and Blue Print	32
3.3	Overall Designed Circuit Diagram	33
3.4	Cross over Design	34

CHAPTER FOUR

ANALYSIS AND DISCUSSION 35

4.1	Signal Coupling and Pre-Amplifier	35
4.2	Multiple Feedback (MFB) Band-Pass Circuit	35
4.3	The Band Pass Design Calculated Values	36
4.4	Output Section and Components Selection	38
4.4.1	Capacitor Selection	39
4.4.2	Component Valves	40
4.5	Peak Limiter/ Attenuator	40

CHAPTER FIVE

RESULTS AND CONCLUSION 42

5.1	Filter Shapes, Responses and Combining Action	42
5.2	Noise Analysis	49

REFERENCES AND BIBLIOGRAPHY 52

ABSTRACT

The versatility of signal analyzing tools: Laplace transform, Fourier series, D.C. and transient analysis were the glaring tools or methods used to ascertain the realization of all constructed circuit here; 'Casio fx7400G' language was used during programming of the band pass filter used quite interesting is to also see the practicability of some engineering theories the reality of signal audio frequencies and their response in tone control. Active filters were used through. This adds a lot of cost, efficiency, size compact and better amplification advantages over the inductive (passive) type. Represented in this critique are fundamental of filters, superiority of active filters over passive filter, overview of Chebychev, Ideal Butterworth, Chebyshev, and Bessel in chapter one whole chapter two covers review of transfer function, Thomas Biquads, Akerbergmossberg Biquard, positive Gain single Amplifier Biquads, Negative Gain single Amplifier Biquads parallel active filter, leapfrog filter, band pass architectures, stages type and high orders that are available. Chapter three discussed multiple feedback topologies the design overall circuit diagram, MFB bandpass computation, crossover design, combining performance, filter responses and combining action. The chapters closed with simulated waveform or filter shapes, responses and combining action noise analyses the audio multi-tone refiner technical specification and a section on practical suggestion (chapter five) and an insight for future research. The desired results were achieved to an extent: the circuit design of an audio multi-tone refiner with a ballpark good quality of production that gave room for frequency bands selection (as enviable) within the audio frequency range; the scientific designing tools or methods as stated above were all realistic in frequency dependent electronic circuit(s) design. Here within is intelligence of a designed and constructed audio device, awfully zest but not with case.

FOREWORD

This work was a research work in all its facets that involved laboratory findings typically designed and constructed; with rigorous mathematical analysis, programming, circuit synthesis and stimulation; bearing in mind the necessities to add to the world of audio production.

CHAPTER ONE

1. INTRODUCTION

The dispensable needs of alignment and optimization of sound systems has prompted this project. In concise answer would be provided to question such as what constituted audio and the frequency response components. Are they pragmatic? What are filter and the rest?

The main focus of this work was to improve an input audio signal to be free from shape glitch in order to circumvent unwanted audible peaks or anomalies in the final sound and offer operational flexibility hence the definition of an Audio Multitone Refiner.

1.1 What Is Filter?

A filter is a device that passes electric signals at certain frequencies of frequency ranges while preventing the passage of other. Filter circuits are used in a wide variety of applications. In the field of telecommunication, band-pass filters are used in audio frequency range (0 kHz to 20 kHz) for modems and speech processing. High frequency band-pass filters (several hundred MHz) are used for channel selection in telephone central offices. Data acquisition systems usually require anti-liaising low-pass filter as well as low-pass noise filters in their preceding signal conditioning stages. System power supplies often use band-rejection filters to suppress the 50 or 60-H line frequency and high frequency transients. In addition, there are filters that do not filter any frequencies of a complex input signal, but just add a linear phase shift to each frequency component, thus contributing to a constant time delay. These are called all-pass filters. At high frequencies (> 1 MHz), all of these filters usually consist of passive components such as inductors (L), resistors (R), and capacitors (C). They are then called LRC filters. In the lower frequency range (1 Hz to 1 MHz), however, the inductor value becomes very large and the inductor itself gets quite bulky, making economical production difficult. In these cases, active filters become important. Active filters are circuits that use an operational amplifier (op amp) as the active device in combination with some resistors and capacitors to provide an LRC-like filter performance at low frequencies (Fig. 1a & b). Inductor based circuits are heavy, expensive to produce and suffer from low frequency distortion and induced hum.

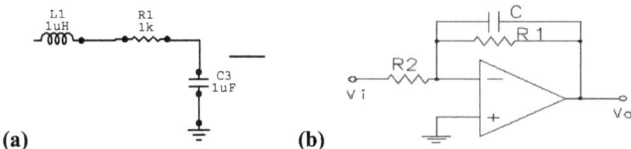

Figure 1. Second-order passive low-pass and second-order active low-pass

Inductors have always been a problem in electronics (audio), as they are by nature relatively large, tends to pick up mains hum as well as other noise in the electromagnetic spectrum. An inductor could be simulated and placed on chips; nevertheless, it has limited Q unlike the real inductor, but very high Q is rarely needed in audio. $L=C_1R_1(R_2-R_1)$ most pronounced limitation are that one end of the simulated inductor is earthed, floating ones are expensive and rare; R_1 minimum allowed is 100ohm (series resistance equivalent to wire "real inductor" resistance), it does not have the same energy storage. The simulated inductor will still try to meet real properties but quit not the same.

The operational amplifier was initially a vacuum tube circuit used in the early 1940s in analog computers. It was so called because it could be used as a high gain dc "amplifier" performing mathematical "operations". Op-Amp's internal circuit is simply a combination of class A, class B and class AB amplifiers. These circuits function together to give a very high gain output.

1.2 Fundamentals of Filters (Low-Pass)

The simplest filter is the passive RC low-pass network where the complex frequency variable allows for any time variable signals. For pure sine waves, the damping constant becomes zero; for a normalized presentation of the transfer function, s is referred to the filter's corner frequency, or -3 dB frequencies. With the corner of the low fc = 1/2RC, for frequencies w >> 1, the roll off is 20 dB/decade. For a steeper roll off, n filter stages can be connected in series to avoid loading effects, op amps, operating as impedance converters, separate the individual filter stages.

Passive band-pass filter involve all the well-known problems of using inductors. In addition, if constant-Q behavior is to be maintained, their outputs require buffering from the loading effects of the slider. Gyrators may be substituted for the inductors; however, at least two operational amplifiers per section will be required (Bohn, 1976). This brings up the next category of active tow-pole RC filters. Requiring only one operational amplifier per section, they represent the most cost-effective approach. While this category of filters is more sensitive to component tolerances than the state-variable approach, the cost advantages are overwhelming. For significantly less money than the cost of the additional two or three operational amplifiers, some very precise passive components can be bought, with precision parts.

Selection from among the various configurations of active RC two-pole bandpass filters is no easy task. Two circuits, however, emerge as time-tested and worthy of further study. Both have been derived from the monumental work of Sallen and Key (14). The first is the well-known voltage-controlled voltage source (VCVS) bandpass filter credited to Kerwin and Huelsman (15) and Multiple feedback Band pass (MFB). It is the most popular non-inverting configuration and features a low spread of element values. A definite advantage is the ability to precisely set the gain of the filter with resistors without upsetting the center frequency. This circuit drops right into BP block shown in Fig. 2.

1.3   Superiority of Active Filters over Passive Filters

Op-Amp provides gain hence the input signal passed to the output will not be attenuated, and therefore better response curves can be obtained. The high input impedance and low input impedance of the Op-Amp means that the filter circuit does not interfere with the signal source or lad. Also, because active filters provide gain, resistor can be used instead of inductors and therefore active filters are generally less expensive. Signal is not affected as much as with passive crossovers, since everything is done in low voltage. There is much more flexibility since all that is needed to adjust crossover frequencies is to turn a knob, while on passive, the component have to be replaced. The problem is that more amplifier channels are needed to go to all the speakers. Passive crossovers work after the amplifiers, receiving high signal cutoff frequencies and filter dB/octave simile.

Cut-off frequency is quite different from octave frequencies. The octave or decade frequencies are basically used for filter topologies, stages or orders contract or definition. Slope or gain drop at octave depends or varies from filter (order) to filters dissipated in Tables. At octave or decade is the standard "checkpoint" frequency to examine attention (or boost) incur on signal frequencies by various filter orders.

Cut-off frequency (wc_1 or wc_2) is not defined by octave or decade frequencies 'scale' since cut-off frequencies is expected to be less than that octave frequency point; at cut off speaker receives half power as compared to the maximum possible power at the center frequency in the circuit. Since half power different in audio amount to about -3dB power which is the maximum tolerable change for human audible frequencies, if the dB loss is greater than or equal to 3dB than this will be clear and wide sparity of sound of different level and to make a quick or faster 'dieing' of the unwanted frequencies attenuation beyond 3dB equivalent frequency has to be more tense has the beauty of the higher order filter is to provide strong more effective 'killing' of the undesired frequencies.

The passive filters have high signal levels since all the frequency splitting is done after the amplifier channel obtaining maximum power by playing with the resistances 'seen' by the amplifier and since an inductor stores current while capacitor stores voltage they act as short and open at low frequencies relatively. The more component added, he more effectively the filter would be. The capacitors and inductors also dissipate power, wasting energy that speakers could be using; inductors are more expensive than capacitor so that passive crossover can get really expensive especially at low frequency of high power applications. Passive also introduce phase shifts (which is often ignored for practical purposes) which put voltage and current out of phase with respect to each other hence affecting the delivered power to the speaker and affecting the delivered power to the speaker and affecting the delivered power to the speaker nad affecting overall speaker 'timing'. A 6dB/octave (which is either a series capacitor or a series inductor filter) has a phase shift of 90 degrees; 12dB/octave (which resulted from a series inductor-capacitor filter) gives 270 degrees phase shift. Anyway! Even order crossover filter would or should be adhered to since this hock up the speakers out of phase (+ to – and – to +).

1.4 Basic Op-Amp and Feedback Theory

No current flows into or out of the input terminals and when negative feedback is applied the differential input voltage is reduced to zero. The voltage follower is extremely useful for buffering voltage sources and for impedance transformation. The impedances of the two inputs should be equal to reduces offsets due to bias currents.

Table 1.1 Filter order and their rate of attenuation (in dB)

ORDER	SECTION	NOMINAL ROLLOFF
First	1	6db/ octave
Second	2	12db/ octave
Third	3	18 db / octave
Fourth	4	24db /octave
"n"	n>4	6db/octave /section

In audio, the first four are common since the transient response become worse and phase for higher orders. Low and high pass filters are usually conventional enough but band pass and band stop filter can be made in many different ways hence a few basic filter alignments (Table 1.1). Butterworth is common in audio while Chebyshev alignment is very common in acoustical filters but it is not generally considered desirable in electronic filter for crossover or other purposes (Franko, 1988).

Speech fundamental as dealt with in this project occur over a fairly limited range between about 125HZ and 250HZ. Vowels essentially contain the maximum energy and power of the voice, occurring over the range of 350HZ to 2000HZ. Consonants occurring over the range of 1500HZ to 4000Hz contain little energy but are essential to intelligibility. The frequency range from 63 to 500Hz carries 60% of the power of the voice and yet contributes only 5% to the intelligibility. The 500Hz to 1 kHz region produces 35% of the intelligibility, while then from 1 to 8 kHz produces just 5% of the power of 60% of the intelligibility, by rolling off the low frequencies and accentuating the range from 1 to 5 kHz, (see chapter three).

1.5 Overview: Block Diagram

Given this design, flexibility was required in the power supply circuitry so that a consistent +/- 30VDC was provided to the amplifier regardless of power input. Previous to the power amplification stage, the audio signal is passed through a high input impedance buffer, multitone refiner, volume/balance control, and a peak limiter protection circuit.

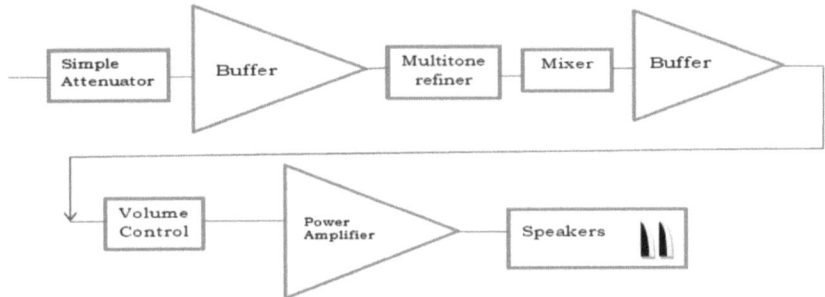

Figure 2: *Block Diagram of an Audio Multitone Refiner*

The preamp section of the amplifier was made up of the Buffer and Peak Limiter. It conditioned the audio signal before the Power Amplification occurred. The Buffer circuit is used so that the current that was drawn from the source (player) was not large. Between the input stage and output stage was multiple feedback bandpass filter that were driven from an Op-Amp buffer which now acts as a low impedance source in the MFB 'bank' input.

The Audio Multitone Refiner was made up of 23 stages and its combination made up of an active circuit that uses a capacitor. The volume control was just a simple circuit while peak limiter (attenuator) works in such a way that when the voltage on the input exceeds it locks the maximum feedback voltage at that value causing the gain to decrease as the input voltage increases.

Most audio would require Q of about 4 which suited to a 1/3 octave filter set. A filter with Q of 10 is too high for audio applications.

1.6 Terms Defiition

The input stage ('front end'): this is to receive and prepare the input signals for 'amplification' by the output stage; class A (with low distortion) is most used here which is built within the Op-Amp.

Output stage: the output stage is the portion which actually converts the weak input signal in to a much more powerful 'replica' which is capable of driving high power to a speaker.

A buffer: it is a circuit that will duplicate the input voltage but provide drive.

CHAPTER TWO

2. LITERATURE REVIEW

Active device (operational amplifier op amp) supersedes inductor in audio processing industries; conversely op amp was at the outset a vacuum tube circuit used in the early 1940 in anlog computer. It was so called because it could be used as a high gain dc amplifier performing mathematical operation op amp's internal circuit is simply a combination of class A class B and class AB amplifiers (Franko, 1988). These circuit functions together to give a very high output gain.

Inductors have always been a problem electronic audio however as they are by nature relatively large tend to pick up mains hum as well as other noise in the electromagnetic spectrum. An inductor is stimulated to be placed on chip; nevertheless, it has limited Q unlike the real inductor but very high Q is rarely needed in audio. With gyrator, the valve of an inductor could be calculated by: L=R1R2C1, more accurately (due to Seigfried linkwitz) is $L=C_1R_1(R_2-R_1)$ most pronounced limitation are that one end of the simulated inductor is earthed, floating ones are expensive and rare; R1 minimum allowed is 100ohm (series resistance equivalent to wire real inductor resistance), it does not have the same energy storage. The stimulated inductor will still try to meet real properties but quit not the same. L along with C1make up a turned circuit which is connected to the pot so that the turned circuit can be moved from the positive to negative input to vary the gain at the turned frequency. The frequency that the circuit is tuned for can be calculated by:

$$f = \frac{1}{2\pi\sqrt{R5R6C1C2}} = \frac{1}{2\pi\sqrt{LC1}} \qquad (R_1=R_5, R_2=R_6)$$

Instead of using MFB filter the more common gyrator tuned circuit can be used. Cauer (elliptical) filter exhibit equiripple characteristic in both the passband and the stopband. Phase information may be gleaned from the transfer functions by separating them in to real and imaginary parts and then using the relationship:

$$\text{Phase: } o = \tan^{-1}\frac{Im}{Re}$$

Group delay is defined as the negative of the first derivative of the phase with respect to frequency.

Type Properties:

Butterworth

- Maximally flat near the center of the band.
- Smooth transition from pass to stopband.
- Moderate out of band rejection
- Low group delay variation near center of band.
- Moderate group delay variation near band edges
- Table of poles for N=1 to 10
- Butterworth has monotonic amplitude response with a maximally flat passband, less phase shift, and better transient results; in conclusion,
- It is the preferred choice (in this project it was only be used in cross over design which was not constructed due to time and financial constrain).

Chebychev
- Equiripple in passband
- Abrupt transition from passband to stopband.
- High out of band rejection.
- Rippled group delay near center of band
- Large group delay variation near band edges.
- Table of poles for N=1 to 10.
- The cherbyshev approximation to an ideal filter has a much more rectangular response in the region near cutoff than has the Butterworth family.

Bessel

- Rounded amplitude in passband
- Gradual transition from passband to stopband
- Low out of band rejection
- Very flat group delay near center of band.
- Flat group delay variation near band edges
- Tables of poles for N= 1to 10

Ideal

- Flat in the passband.
- Step function transition from passband to stopband.
- Infinite out of band rejection.
- Zero group delay everywhere.

Outline

Butterworth	Advantages maximally flat magnitude response in the passband. Good all round performance pulse response better than chebyshev. Rate of attenuation better than Bessel.	Disadvantages some overshoot and. Ringing in step response.
Chebyshev	Advantages better attenuation Beyond the pass band than Butterworth pass-band than Butterworth advantages best step response	Disadvantages ripple in pass band. Considerable ringing in step response
Bessel	Advantages best step response very little overshoot or ringing	Disadvantages slower rate of attenuation beyond the pass band than Butterworth

2.1 Transfer Functions

Transfer function is created in three forms, standard, cascade, and parallel. Cascade and parallel transfer functions consist of first and second order terms that are cascaded or summed in parallel together. The cascade and parallel transfer are used to create the active filters. Cascade transfer function generate the filter composed of Thomas biquads, positive gain single amplifier biquads, negative gain single amplifier, biquads and GIC biquads, parallel transfer function are implemented with a summation of positive gain and negative gain single amplifier biguads.

Typical transfer functions are below:

$$\frac{S(S2)}{S2 + S + 2)(S + 4)}$$

Cascade transfer function

$$1 + \frac{S}{S2+2S+2")(S+4} + \frac{4}{S+3}$$

Parallel Transfer function

Thomas Biquads
Their primary advantage is that it provides very high Q second order stages.

Akerberg-Mossberg Biquads
The Akerberg-Mossberg biquad exceeds the performance of the Thomas biquads for Opamp imperfections and matches the Thomas 2 biquad notch performance in the presence of element valve errors. This increased performance is obtained by replacing the positive integrator in the Thomas 2 biquad second and third Opamps with a Miller integrator. The Miller integrator uses

two matched op amp in a configuration that tends to cancel errors due to Opamp imperfections. The Akerberg-Mossberg biquad may absorb a third pole.

Positive Gain Single Amplifier Biquads
"The advantages of positive Gain SAB'S, except for Twin T stages are always gain changeable, and there is usually no reversal of sign. The disadvantages are they are more susceptible to element imperfections than negative Gain SAB's. All pass and even notch positive Gain SAB's with injector resistor have assigned reversal"[16].

Negative Gain Single Amplifier Biquads
Negative gain single amplifier biquads (SAB's) require only one op amp for first, second, and sometimes third order amplifiers. Second and third order Negative Gain SABs use the Bridge T or MFB circuit configuration for the feedback path. Filter light does not support third order stages. "The advantages of Negatives Gain SAB;s is that they are generally higher resistant to imperfect elements than positive gain SAB's. The disadvantages is that are generally more susceptible to op amp imperfections than positive gain SAB's and the gain for all pass notch stages is fixed"[17].

Parallel Active Filters
The summation circuit may be optionally active or passive. The advantage of doing this is that performance degradation due to op amp imperfections is not amplified through successive cascade stages. The disadvantage is the filter design may be physically very large and notches tend to be poor quality.

Leapfrog Filters
Leapfrog filter are passive LC ladder simulations. The advantage is that errors due to element values or op amps tend to be distributed across the filter instead of concentrated at a specific biquad. This generally makes them more robust. Filter solutions supports leapfrog filters for low pass and band pass all pole designs; Table 2.1 illustrate quick view of the above discusses stages.

Band Pass Architectures
Band pass filter may be created with multiple integrated band pass stages, or high and low pass stages. Odd order filters of the high/ low pass architecture always have a band pass stages in the center. In general, the integrated band pass architecture works better for narrow band filters, and the high/ low pass architecture works better for wide band filters.
This is due to potentially huge, undesirable internal gains that may saturate op amps if the wrong architecture is used.

Stages Type	High Orders That Are Available
Thomas or Akerberg –Mossberg Biquads	Third Order
Positive gain all poles single amplifier stages	Third order and fourth order
Positive gain single amplifier stages with transmission zeros	Third order
Negatives gain all poles single amplifier stages	Third order and fourth order
Negatives Gain Single Amplifier Stages With Tranmissinon zeros	Usually Third Order
GIC	Third Order
Twin T	Third Order In The Form Of An Rc Pole Following the Op Amp

Fundamentals of low pass filters
The simplest low pass filter is the passive RC low pass network shown in Fig. 3

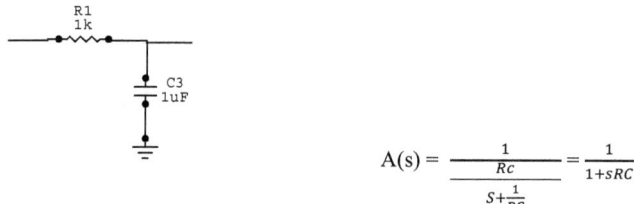

$$A(s) = \frac{\frac{1}{Rc}}{S+\frac{1}{RC}} = \frac{1}{1+sRC}$$

Figure 3. *First order passive RC Low-Pass and its transfer function*

Where the complex frequency variable, s=jw+0, allows for any time variable signals; for pure sine waves, the damping constant, 0, becomes zero and $^{s=jw}$.
For a normalized presentation of the transfer function, s is referred to the filter's corner frequency, or -3 dB Frequency, ^wc, and has these relationship:

$$s = \frac{S}{wc} = \frac{jw}{wc} = j\frac{f}{fc} = jw$$

With the corner frequency of the low- pass in Figure 2.1 being fc=1/wRC, s becomes S= sRC and the transfer function A(s) result in:

$$A(S) = \frac{1}{1+s}$$

The magnitude of the gain response is $|A| = \frac{1}{\sqrt{1+\pi 2}}$

The frequencies n>>1, the rolloff is 20 Db/decade. For a steeper rolloff, 'n' filter stages can be connected (James, 1999) in series as shown in Fig.4. To avoid loading effects, op amps, operating as impedance converters, separate the individual filter stages.

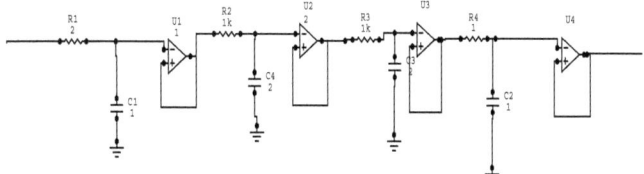

Figure 4. Fourth – order passive RC Low pass with Decoupling Amplifiers

The resulting transfer function is:

$$A_{(S)} = \frac{1}{(1+\alpha 1S)(1+\alpha 2S)...(1+\alpha nS)}$$

In the case that all filters have the same cut-off frequency, fc, the coefficients become a1=a2=...an=a=$\sqrt{2}-1$ and fc of each partial fiter is 1/α times higher than fc.

Fig. 3 shows the results of a fourth order RC low-pass filter. The roll off of each partial fiter(curve 1) is- 20 dB/ decade, increasing the roll-off of the overall filter (curve 2) to 80 dB/ \decade.

2.2 Butterworth Low Pass Filters

The Butterworth low pass filter provides maximum pass band flatness. Therefore, a Butterworth low pass is often used as anti-aliasing filter in data converter applications where precise signal levels are required across the entire pass band. Fig. 4 plots the gain response of different orders of Butterworth low pass filters versus the normalized frequency axis Ω(Ω=f/fc); the higher the order, the longer the passband flatness

2.3 Tschebyscheff Low Pass Filters

The Tschebyscheff low pass pass filters provide an even gain rolloff above Fc; however, the passband gain is not monotony, but contains ripples of constant magnitude instead. For a given filter order, the higher the passband ripples, the higher the filter's rolloff. With increasing filter order, the influence of the ripple magnitude on the filter rolloff diminishes. Each ripple accounts for one second-order filters with order numbers generate ripples above the 0-dB line, while filters with odd numbers create ripple below 0 dB. Tschebyscheff filters are often banks, where the frequency content of a signal is of more important than a constant amplification.

2.4 Bessel Low Pass Filters

The Bessel low pass filters have a linear phase response over a wide frequency range, which results in a constant group delay in that frequency range. Bessel low pass filters, therefore, provide optimum square- wave transmission behavior. However, the passband gain of a Bessel low pass filter is not as flat as that of the Butterworth low pass, and the transition from passband to stop band is by far not a sharp as that of a Tschebyscheff low pass filter.

2.5 Quality Factor Q

The quality factor Q is an equivalent design parameter to the filter order n, instead of designing an nth order Tschebyscheff low pass; the problem can be expressed as designing a Tschebyscheff low pass filter with a certain Q. the band pass filter, Q is defined as the ratio of the mid frequency, f m, to the bandwidth at the two -3dB points:

$$Q \frac{Fm}{(f2-f1")}$$

For low pass and high pass filters, Q represents the pole quality and is defined as

$$Q: \frac{\sqrt{b1}}{a1}$$

High Qs can be graphically presented as the distance the 0-dB line and the peak point of the filter's gain response.

2.6 Précis

The general transfer function of a low pass filter is:

$$A(s) = \frac{A0}{\Pi(1+a1s+b1s)}$$

The filter coefficients ai and bi distinguish between Butterworth, Tschebyscheff, and Besssel filters Tab book inc., 1982). The coefficients for all three types of filters could be found in filter Table; in addition, the ratio $\frac{\sqrt{bi}}{ai=Q}$ is defined as the pole quality. The higher the Q valve, the more a filter inclines to instability.

2.7 Low Pass Filter Design

The transfer function of single stage is:

$$A(s) = \frac{A0}{(1+a1s+b1s2)}$$

For a first order filter the coefficient b is always zero (b1=0), thus yielding:

$$A(s) = \frac{Ao}{1+a1s}$$

The first order and second order filter stages are the building blocks for higher-order filters. Often the filter operate at unity gain (A O=1) to lessen the stringent demands on the op amp's open loop gain. Figure 2.8 shows the cascading of filter stages up to the sixth order. A filter with an even order number include number include an additional first order stage at the beginning.

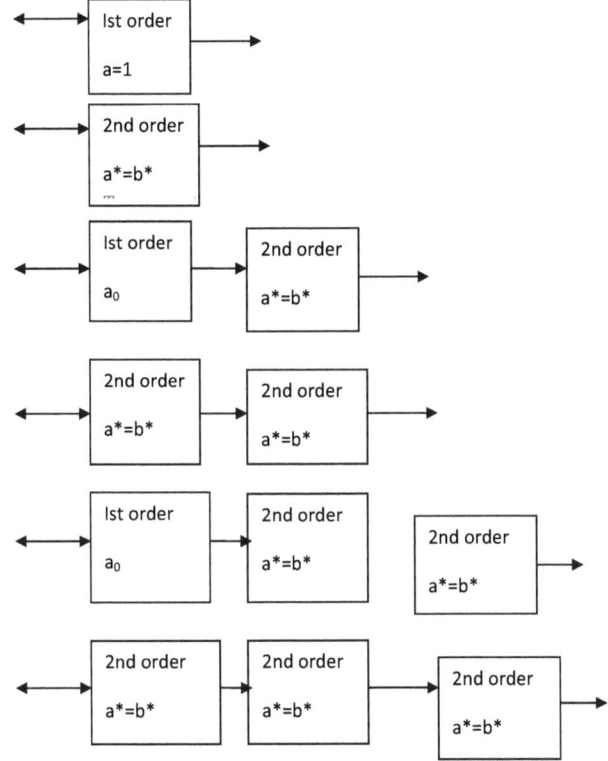

Figure 5. Cascading filter stages for higher order filters

To avoid the saturation of the individual stages, the filter needs to be placed in the order of rising Q valves.

2.8   First Order Low Pass Filter

Fig. 6a and 6b show a first order low pass filter in the inverting and in the non-inverting configuration.

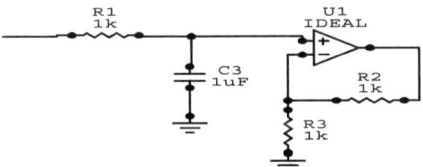

Figure 6a. First order non inverting low pass filter

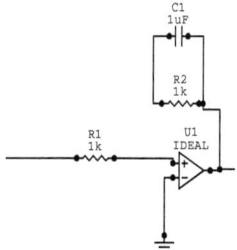

Figure 6b. *First order inverting low pass filter*

The transfer functions of the circuit are:

$$A(s) = \frac{1 + \frac{R2}{R3}}{1 + W_c R1 C1 S} \quad \text{and} \quad A(s) = \frac{-\frac{R2}{R3}}{1 + W_c R1 C1 S}$$

The negative sign indicates that the inverting amplifier generates a 180- degree phase shift from the filter input to the output[6]. The coefficient comparison between the two transfer function and the previous equation yields:

$$A_0 = 1 + \frac{R2}{R3} \quad \text{and} \quad A_0 = -\frac{R2}{R3}$$

$$a_1 = w_c R_1 C_1 \quad \text{and} \quad a_1 = w c R_2 C_1$$

To dimension the circuit, specify the corner frequency (fc), the dc qain (Ao), and capacitor C_1, and then solve for resistor R_1 and R_2:

$$R_1 = \frac{a1}{2\pi f c C1} \quad \text{and} \quad R_2 = \frac{a1}{2\pi f c C1}$$

$$R_2 = R_3 (Ao-1) \quad \text{and} \quad R_1 \frac{-R2}{Ao}$$

The coefficient a_1 is taken from one of the coefficient Tables, that all filter types are identical in their first order and $a_1 = 1$. For higher filter, however, $a_1 \neq 1$ because the corner frequency of the first order stage is different from the corner frequency of the overall filter. When operating at unity gain, the non-inverting amplifier reduces to a voltage follower thus inherently providing superior gain accuracy. In the case of the inverting amplifier, the accuracy of the unity gain depends on the tolerances of the two resistors R_1 and R_2

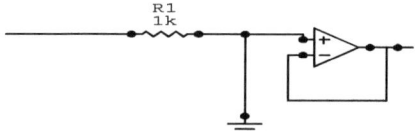

Figure 7a. *First order Non-inverting low pass filter with unity Gain*

The two topologies for a second-order; low pass filter, the Sallen-key and the Multiple feedback (MFB) topology will be reviewed next.

2.9 Sallen-key Topology

The general Sallen-key topology in Fig.7a allows for separate gain setting via Ao=1+(R_4/R_3). However, the unity-gain topology in Fig.7b is usually applied in filter designs with gain accuracy, unity gain, and low Qs (Q<3).

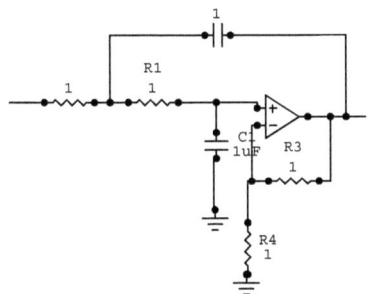

Figure 7b. General Sallen –Key Low-Pass Filter

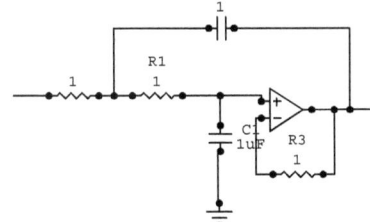

Figure 8. Unity gain Sallen-key low pass filter

The transfers function of the circuit in Fig.7a is

$$A_{(s)} = \frac{Ao}{1+wc[c1(R1+R2)+(1-Ao)R1C2]s+wc2R1R2C1C2S2}$$

For the unity gain circuit in Figure 2.11b (Ao=1), the transfer function simplifies to:

$$A_{(s)} = \frac{1}{1+\omega cC1(R1+R2)S+\omega c2R1+R1R2C1C2s2}$$

The coefficient comparison between this transfer function yields:

$$Ao=1$$
$$a_1 = wcC_1(R_1+R_2)$$
$$b_i = wc^2 R_1 R_2 C_1 C_2$$

Given C_1 and C_2, The resistor valves for R_1 and R_2 are calculated through:

$$R_{1,2} = a_1c_2 = \frac{\sqrt{a1c2} - 4b_1C_1}{4\pi fcC1C2}$$

In order to obtain real valves under the squre root must satisfy the following conditions

$$C_2 \geq \frac{C_1 4b1}{a_1^2}$$

A special case of the general Sallen key topology is the application of equal resistor valves and equal capacitor valves: $R_1=R_2=R$ and $C_1=C_2=C$. the general transfer function changes to:

$$A(s) = \frac{Ao}{1+wcRc(3s-Ao)s+(WcRC)22} \quad \text{with} \quad A_0 = 1 + \frac{R4}{R3}$$

The coefficient comparison also yields:

$$a1 = \omega cRC(3-Ao)$$
$$b1 = (\omega cRC)^2$$

Given C and solving for R and Ao results in:

$$R = \sqrt{\frac{b1}{2\pi fc}} \quad \text{and} \quad Ao = 3 - \frac{a1}{\sqrt{b1}} = 3 - \frac{1}{Q}$$

Thus, Ao depends solely on the pole quality Q and vice versa; Q, and with it the filter type, is determined by the gain setting of Ao:

$$Q = \frac{1}{3-Ao}$$

The circuit in Fig.9 allows the filter type to be changed through the various resistor ratios R_4/R_3.

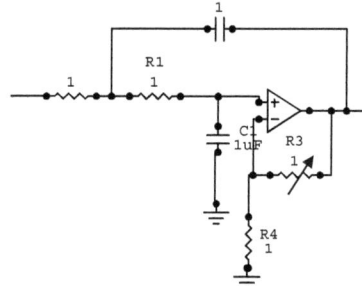

Figure 9. *Adjustable second order low pass filter*

Tables 2.1 lists the coefficients of the second order filter for each filter type and gives the resistor ratios that adjust the Q

Table 2.1: *Seconds- order Filter Coefficients*

Second order	Besel	Butterworth	3-db Tschebyscheff
a_1	1.3617	1.4142	1.005
b_1	0.618	1	1.9305
Q	0.58	0.71	1.3
R_4/R_3	0.288	0.588	0.234

Source ([14])

2.10 Multiple Feedback Topologies

The MFB, Fig 10, topology is commonly used in filters that have high Qs and require a high gain.

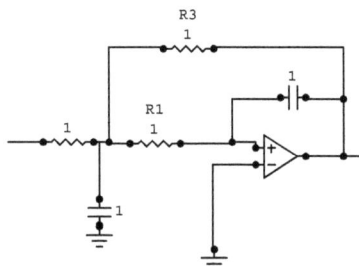

Figure 10. Second-order MFB low pass filter

2.11 Higher Order Low-pass Filters

Higher order low pas filter are required to sharpen a desired filter characteristic. For that purpose, first order and second order filter stages are connected in series, so that the product of the individual frequency responses results in the optimized frequency response of the overall filter. In order to simply the design of the partial filters, the coefficients a_1 and b_1 for each filter are listed in the coefficient Tables.

2.11.1 Second Order Filter

Fig. 11. Illustrated second-order unity-Gain Sallen=Key Low-Pass Filter.

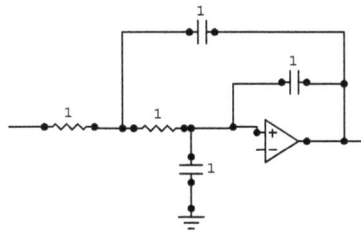

Figure 11. Second-order unity-Gain Sallen=Key Low-Pass Filter

$$C2 \geq C1\frac{4b2}{a2;} \quad R1 = \frac{a_2C_2 - \sqrt{a2C2 - 4b2c1c2}}{4\Pi f_c C_1 C_2}$$

And $R_2 = \dfrac{a_2 C_2 - \sqrt{a2C2 - 4b2c1c2}}{4\Pi f_c C_1 C_2}$

2.11.2 Third Order Filter

The calculation of the third filter is identical to the calculation of the second filter, except that a_2 and b_2 are replaced by a_3 and b_3, thus resulting in different capacitor and resistor valves. Fig. 12 shows the final filter circuit with its partial filter stages.

$C_2 \geq C_1 \dfrac{4b3}{a32}$

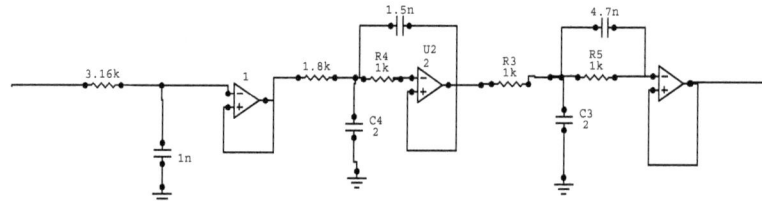

Figure 12. *Fifth order unity Gain Butterworth Low pass filter*

2.12 High Pass Filter Design

By replacing the resistor of the low pass filter with capacitors, and its capacitors with resistors, a high pass filter is created, see Fig. 13.

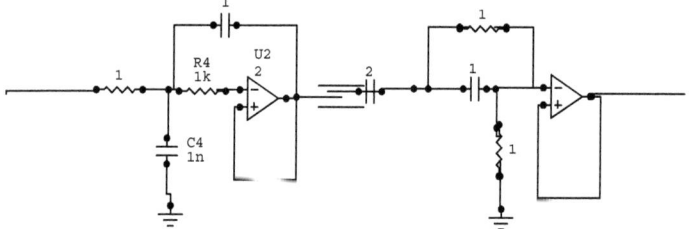

Figure 13. *Low Pass Filter Transition through Component Exchange*

The general transfer function of the high pass filter is:

$$A_{(s)} = \dfrac{A}{\pi(1+\dfrac{ai}{s}+\dfrac{bi}{s})}$$

With $^A\omega$ being the pass-band gain, filter coefficient could be determined from the Table in appendix A.

2.13 Multiple Feedback Topologies

The MFB topology is commonly used in filters that have Qs and require a high gain. To simply the computation of the circuit, capacitors C1 andC2 assume the same valve (C1=C3=C). The transfer function of the circuit is

$$A(S) = -\frac{\frac{C}{C2}}{1+\frac{2C+C2}{\omega CR1CC2}\cdot\frac{1}{s}+\frac{2C+C2}{wCR1CC2}\cdot\frac{1}{s^2}}$$

Through coefficient comparison with equation above, obtain the following relations:

$$A\omega = \frac{C}{C2}$$
$$a_1 = \frac{2C+C2}{\omega cR1CC2} \quad R_1 = \frac{1-2A\alpha}{2\pi f.C.a1}$$
$$b_1 = \frac{2C+C2}{\omega cR1CC2} \quad R_2 = \frac{a1}{2fc.b1C2(1-2A\alpha)}$$

Given capacitors C and C2, are solving for resistors R1 and R2. The pass-band gain A(∞) of a MFB high pass filter can very significantly due to the wide tolerances of the two capacitors Cand C2. To keep the gain variation at a minimum, it is necessary to use capacitors with tolerance valves.

2.14 Band-Pass Filter Design

In section the previous section, replacing the term S in the low pass generated a high pass response transfer function with the transformation I/S. likewise, a band pass characteristics is generated by replacing the S term with the transformation

$$\frac{1}{\Delta\Omega}\left(s+\frac{1}{s}\right)$$

In this case, the pass band characteristics of a low pass filter is transformed into the upper passband half of a band filters, Fig.14. The upper pasband in then mirrored at the mid frequency, $fm=1$, into the lower passband half.

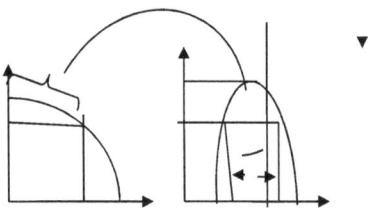

Figure 14. Low pass to band transition

The corner frequency of the low pass filter transforms to the lower and upper 3Db frequencies is defined as the normalized band widthΩ

$$\Delta\Omega = \Omega 2 - \Omega 1$$

The normalized mid frequency, where Q=1, is:

$$\Omega_m = 1\Omega_2\Omega_1$$

In analogy to the resonant circuit, the quality factor Q is defined as the ratio of the mid frequency (fm) to bandwidth (B):

$$Q = \frac{fm}{B} = \frac{fm}{f2-f1} = \frac{1}{\Omega 2 - \Omega 1} = \frac{1}{\Delta\Omega}$$

The simplest design of a band pass filter is the connection of a high pass filter and a low pass filter in series, which is commonly done in wide band filter application. Thus, a first order high pass and first order low pass provide a second order band pass, while a second order high pass and a second order low pass result in a fourth order ban pass response. In comparison to wide band filters, narrow band filters of higher order consist of cascade second order band pass filter that use the Sallen key or the multiple feedback (MFB) topology.

2.14.1 Second Order Band Pass Filter

To develop the frequency response of a second order band pass filter, apply transformation to a first order low pass transfer function:

$$A(s) = \frac{Ao}{1+s} \quad \text{Replacing } \frac{1}{\Delta\Omega}(s + \frac{1}{s})$$

Yield the general transfer function for a second order band pass filter:

$$A(s) = \frac{Ao\Delta\Omega s}{1+\Delta\Omega s + s}$$

When designing band pass filter, the parameters of interest are the gain at the mid frequency(A_m) and the quality factor (Q), which represents the selectivity of a band pass filter. Therefore, replace Ao with A_m and Ω with 1/Q gives:

$$A(s) = \frac{Am}{Q}S$$

Fig.15 show the normalized gain response of a second order band pass filter for different Qs.

2.14.2 Sallen Key Topology

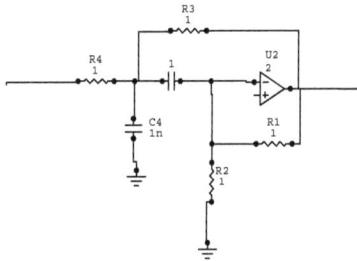

Figure 15. Sallen-key Band-Pass

The Sallen-key band-pass circuit in Fig.15 has the following transfer function:

$$A(s) = \frac{G \cdot RC\omega ms}{1+RC\omega m(3-G)\cdot s + RC\omega ms}$$

Through coefficient comparison obtained the following equation:

Mid frequency: $f_m = \frac{1}{2\pi RC}$

Inner gain: $G = 1 + \frac{R2}{R1}$

Gain at f_m: $A_m = \frac{G}{3-G}$

Filter quality: $Q = \frac{1}{3-G}$

The Sallen-key circuit has the advantages that quality factor (Q) can be varied via the inner gain (G) without modifying the mid frequency (f_m). A drawback is, however, that Q and A m cannot be adjusted independently; Care must be taken when G approaches the valve of three, because then A_m becomes infinite and causes the circuit to oscillate. To set the mid frequency of the band pass, specify fm and C and the R is solved for

$$R = \frac{1}{2\pi fmC}$$

Because of the dependency between Q and Am, there are two options to solve for R_2: either to set the gain at mid frequency:

$$R_2 = \frac{2Am-1}{1+Am} \quad ; \quad R_2 = \frac{2Q-1}{Q}$$

Or to design for a specified Q.

2.14.3 Multiple Feedback Topology

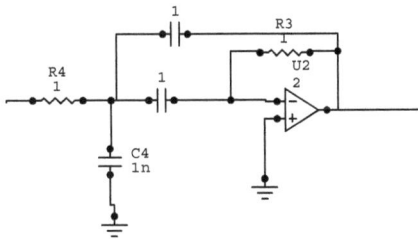

Figure 16. MFB band pass

The MFB band pass circuit in Fig. 16 has the following transfer function:

$$A(S) = \frac{\frac{-R2R3}{R1+R3}C\omega_m \cdot S}{1+\frac{2R1R3}{R1+R3}C\omega_m \cdot S + \frac{R1R2R3}{R1+R3}C^2 \cdot \omega_m^2 \cdot S^2}$$

The MFB band pass allows adjusting Q, A_m, and f_m independently. Bandwidth and gain factor do not depend on R_3. Therefore, R_3 can be used to modify the mid frequency without affecting bandwidth, B, or gain A_m. for low values of Q, the filter can work without R_3, however, Q then depends on A_m via: $A_m = 2Q^2$

Others reviewed are:

a) **Fourth Order Band-Pass Filter (Staggered Tuning)**

Bessel				Butterworth				Tschebyscheff			
a1	1.3617			A1	1.4142			a1	1.0650		
b1	0.6180			B1	1.0000			b1	1.9305		
Q	100	10	1	Q	100	10	1	Q	100	10	1
ΔΩ	0.01	0.1	1	ΔΩ	0.01	0.1	1	ΔΩ	0.01	0.1	1
α	1.0032	1.0324	1.438	α	1.0035	1.036	1.4426	α	1.0033	1.0338	1.39

b) **Band Rejection Filter Design**

c) **Active Twin Filter**

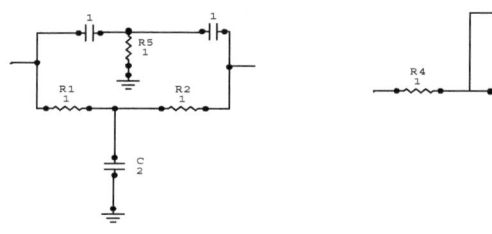

Figure 17. Passive Twin T Filter   *Figure 18. Active Twin T Filter*

The transfer function of the active twin T filter is:

$$A(S) = \frac{K(1+S2)}{1+2(2-K).S+S2}$$

Mid –frequency: $f_m = \frac{1}{2\pi RC}$

Inner gain: $G = 1 + \frac{R2}{R1}$

Passband gain: $A_o = G$

Rejection quality $Q: \frac{1}{2"(2-G)}$

d) **Active Wien Bobinson filter**

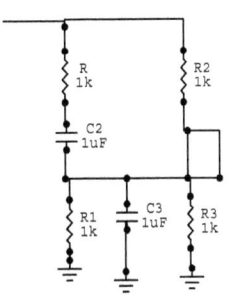 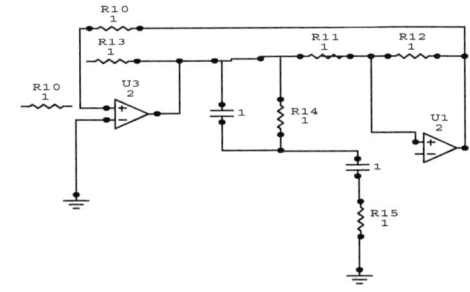

Figure 19. *Passive Wien Robinson Bridge*   **Figure 20.** *Active Wien Robinson Filter*

The active Wien Robinson filter has the transfer function:

$$A(S) = -\frac{\frac{\beta}{1+\alpha}(1+S^2)}{1+\frac{3}{1+\alpha}\cdot S + S^2}$$

with
$$\alpha = \frac{R2}{R3} \quad \text{and} \quad \beta = \frac{R2}{R4}$$

And with comparing the variables, it provides the equations that determines the filter parameter

Mid frequency: $f_m = \frac{1}{2\pi RC}$

Passband gain: $A_o = \frac{\beta}{1+\alpha}$

Rejection quality: $Q = \frac{1+\alpha}{3}$

To calculate the individual component values, the following design procedures are established:

i. define f_m and C and calculate R with:
$$R = \frac{1}{2\pi f_m C}$$

ii. specify Q and determine α via:
$$\alpha = 3Q - 1$$

iii. Specify A_o and determine β via:
$$\beta = -A_o \cdot 3Q$$

iv. Define R_2 and calculate R_3 and R_4 with:
$$R_3 = \frac{R2}{\alpha} \quad \text{and} \quad R_4 = \frac{R2}{\beta}$$

Pass Filter Design

The general transfer function of an all pass is then

$$A(S) = \frac{\pi(1-a_is+b_is^2)}{\pi(1+a_is+b_is^2)}$$

with a_i and b_i being the coefficients of a partial filter

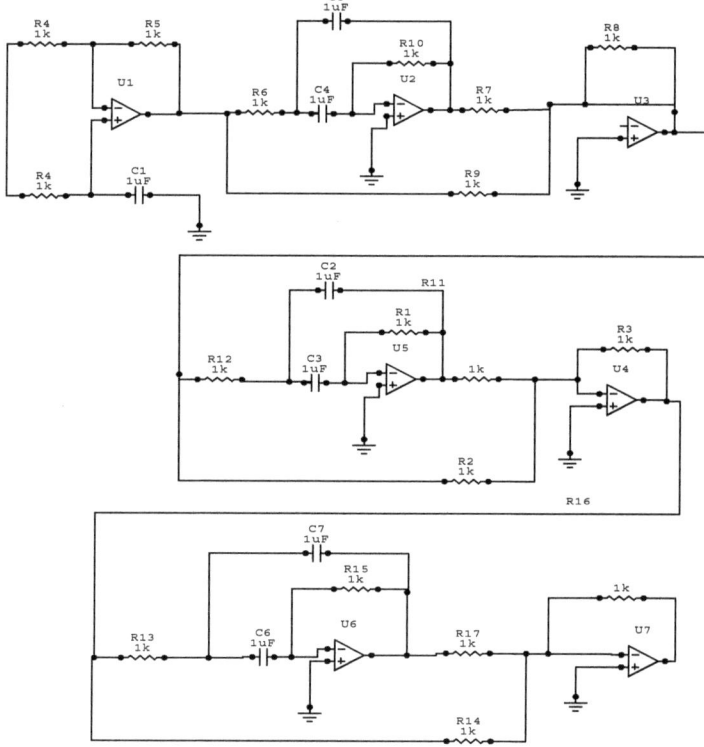

Seven-Order All-Pass Filter

Another type of topology called state- variable (Bohn 1976), band pass circuits have the best performance for the cost is no object designs; other filter such as transverse filter (which main advantage that it offers linear phase) and shelving filter had all been used for similar projects. But due economic reasons MFB is opted for band allocation.

2.15 Tone Control (Baxandall) Circuit

This gives independent variation of bass and tremble without switches; it is fully symmetrical, and there is no interaction between the controls. The OP-AMP acts as a buffer. Frequency response is absolutely flat. Its bass is defined at frequency less then resonant frequency is that where the capacitance and inductance is equal. This provides to give more selectively of desired tone. Tone control of audio systems involves altering the flat response in order to attain more low frequencies or more high ones, dependent upon listeners' preference. The circuit produces 20Db (4th order) of bass or treble boost or cut as set by the variable resistance. The response of the circuit is as shown. Baxandall Tone controls: when centered, there is neither loss nor gain, and the op-amp acts as a buffer.

Crossover

It is hard to make a loud speaker that is capable of handling the entire audio spectrum more difficult to make one that does this well hence the needs for crossover. The use of higher other filter allows loudspeakers to be played at the limits of their efficiency. Higher filter can also be beneficial in compensating for natural peaks within the listening environments (vehicles).

Crossover is an essential part of an audio system (sometime ignored); it splits frequency so that each speaker receives a certain range of frequencies to avoid speaker damage and to maintain overall balance. Speakers are designed only to efficiently play in range of frequencies, if the frequencies are played than the speaker will produce distortion, which will eventually destroy it. If a system with subwoofers (20 to 100Hz) and full ranges speakers does not have crossover, then the subs will be playing while the full rage speakers will be playing from 60Hz. There is an "overlap" of frequencies between 50 and 100Hz, yielding to a non- balanced system. These are three types of crossover: high pass, low pass and band pass. The crossover cut (does not block undesired frequencies completely unless it is digitalized) frequencies progressively. The minimum blocking should be 6dB/ octave (for the least filter order) that is the level at the speaker.

Q-Choice

Most audio applications require a maximum Q of about 4.0 is suited for $_{1/3}$ octave filter set Q of 10 is too high to be useful in most audio application. Bandwidth is measured at the -3dB frequencies on either sides of the resonant peak (ωo). $\omega a \times \omega_{c2:}$ increasing Q does nothing to the roll off slope. The higher gain fall rate (slope) for first order is 6dB/ octave no greats! Octave is the building or halving of a frequency octave division: is by stay frequencies., care must be taken to ensure (Table 2.2) that the loading on the input buffer op-amp is not excessive to that low capacitance cause problems due to stay capacitance on the board the result values becomes too high.

Range   20Hz to 20 KHz
           10 Hz to 10KHz
640Hz OR 1000Hz are nominal center for audio

Table 2.2: *Parts selection Guide*

Fmin (Hz)	Fmax (Hz)	Nano Fared
20	80	330
80	300	82
300	1200	22
1200	4800	5.6
4800	20000	1.5

This Table is suitably used as rough quick guide

Generally filter: Butterworth, Chebyshev and other determines how effective blocking undesired frequencies a filter are (Table 1.1). State-variable band pass circuits have an unsurpassed performance for no cost object designs. Filter "order" gives information about how well (or otherwise) a given filter will reject the unwanted frequencies (Table 1.1). In audio, the first four are common since the transience response become worse and phase disturbances become evident as the filter order increased. All filters affect the phase of signal and all filters have some effect on transient response. These are four main responses that are obtainable from filter low; high, band pass and band stop (reject) filters. Low high filters are usually conventional enough but band pass and band stop filter can be made in many different ways hence a few filter alignments. Butterworth is common in audio while Chebyshev alignment is very common in acoustical filters but is not generally considered desirable in electronics filter for crossover or other purposes. Linkwitz-Rilly is a rearrangement of two cascaded second order Butterworth filters and relies on characteristics of the sections (high pass and low pass). Again, there is different circuit arrangements that are commonly used in audio to create any of the filters described above.

Butterworth is imperative if a high impendence circuit needs to drive a low impedance load or many loads. Since + input to the op amp is already a high impedance input this input resistor is almost irrelevant with the exception of possible protection (or filter) of the input to the op amp. The output impedance is very low. A buffer can also be called a "follower". It mimics the input with the low impedances output drive; theoretically zero ohms but usually 47 ohms. Whatever the input sees, it will duplicate it on the output. This can provide extra drive for another circuit loading the input circuit; transistor utilizes maximum voltage transfer more efficiently than maximum power transfer. Hence, for maximum voltage transfer the source impedance must be very small compare to the destination/ load impedance. Z destination must be at least 10 × of that of the source internal impedance. With modern audio circuit matching can actually degrade audio performance.

CHAPTER THREE

3. DESIGN

The approach exploited here was rigorous and experimentation coupled with, mathematical analysis, programming/ computation- to –simply calculations-, circuit synthesis and simulation or modeling (see special comments below) with Butterworth filter response as the favourite standard in this study; however, multiple feedback band pass (MFB) Filter would be used (approximate) unit- gain frequency bandwidth.

Special Comment: the design package used here is limited in applications in that while it allowed some features to be exploited; it disallowed (subscript) others hence Vin (wherever it is seen in this write-up implies) V_i, likewise Va=Va R1=R1, R2=R2, R3=R3, C1=C1, C2=C2, and C3=C3. These were adhered to in the subsequent chapters expect in some cases where circuit models were properly drawn.

The input low frequency response signal needs to be defined. It is done once through C1 and Rin; to prevent a very saggy bottom end to the frequency response that might result using non inverting as the input a single series capacitor C1 and the bias resistor (the equivalent op-amp input resistance) parallel resonant together to limit the passable or allowable feed in frequency. While for the inverting type, there is series resonant between series C1 and a series input resistor R1, 3Db down at 5Hz, and 0.2Db down at 20Hz, which is definitely inside the audio band. The op-amp were so guided to ensure stable dc biasing round all its terminal hence high pass at about (5Hz) were placed the op-amp to steer tract its dc bias while that along the supplies were to maintain stability at high frequency and to filter distortion frequency in the supplies lines. Once the starting frequency 15Hz 'the necessary octave division and quality factor, Q to use for various frequencies bands had been determined, other frequencies could be factor, taken a crucial reference to main vocal frequencies using the expression :

$$f_{o1} \times k\sqrt{2} = f_{o2}$$

Where f_{o1} is the starting center frequency; f_{o1} is the next upper center frequency, K(=Q-1) is the number of division an octave was divided. Hence Q is known! With this, the next illustrated center frequencies range were selected: 20,25,31,40,50,63,80,100,125,160,200,250,315, 400,500,630,800,1000,2000,4000,8000,16000,20000.

3.1 Filter Architecture

The transfer function of MFB filter used was as depicted in the following analysis:

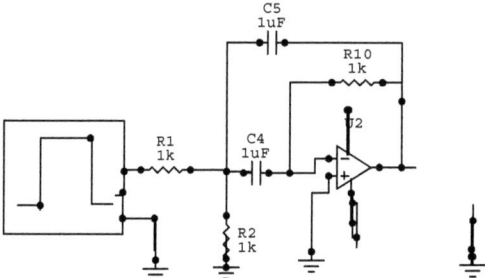

Using KCL, summing current at each node gave

At node a
$$\frac{Va-Vi+(Va-Vo)SC_1 + V_aSC_1=0}{R1} \quad \text{...............3.1}$$

At node b
$$-VaS1+\frac{Vo}{R3}=0 \quad \text{...............3.2}$$

Then
$$H(S)=\frac{Vo}{V1}=\frac{SCR2}{S2(C2R3)+S(CR+2R)}$$

$(R=R_1=R_2=R_3; C_1=C_2)$

Since the standard second order band-pass is
$$H(S)=\frac{KS\beta}{S2+\beta S+\omega 2}$$

$\sqrt{2R} = 2\pi fc$

$fc = \frac{1}{2\pi C} \times \sqrt{2R}$

hence otherwise, $fc = \frac{1}{2\pi C} \times [\frac{1}{\sqrt{R3 \times R1 // R2}}] = \frac{1}{2\pi C} \times \frac{1}{\sqrt{Req \times R3}}$

R_1=input- resistance, R_2=Attenuator- resistance;
R_3 = feedback- resistance

$fc = \frac{1}{2\pi fc} \times \frac{1}{R2} \times \frac{R1+R2}{R1+R3}$

Generally,

when, C1=C2; and $Q/K = R1, R2 = Q/(2Q-K); R_3=2Q$

(Where Q is the center frequency ratio to band frequency)

Design Dimensioning Techniques

The below program simply (see special comments in section 3.0) the component selection having or chosen MFB filter for this design, Table 3.1 shows the resulted valves which are critically chosen to the available component series (E12 and E24). This (program2) caters for various desired frequencies, capacitors and resistors in that order.

Filter program 1 *(Fc calculation only)*
"R1="?→ A;
"R2="?→ B;
"R3=:?→ C;
"Q="? → K;
Lbl1:"C1"? → D:"C2"? → E;
QA÷K → U;
QB÷ (2Q²-K)→V;
2Q^3÷ (2KQ-K²) →W;
((U+V)÷ (4π²*ABCDEW))^O .5→Y;
"FC="Y.
Goto 1;
Program 2:(C1,C2,R1,R2 and R3 can be calculated apiece)
"R1="?→A
"R2="?→B
"R3="?→ C
"C1="?→ D
"C2="?→ E
"Fc=" ?→F
"Q="? →Q
"K="? →K
(2K²Q²-K^3)Q →G;
8F²Kπ²Q^3(Q²-K) →I;
GB+IA→ J;
"What do you want? Input for R1, 2 for R2, 3 for R3,4 for C1 and 5 for C2 X
IF X=1
   Then Goto 1;
      Else if X=3:Then Goto 3;
         If X=4: Then Goto 4;
            Else if X=5:Then Goto 5;
Lbl 1((C) ÷(2K)) →U;
"R1=":U.
Goto 1;
Lbl 2:(1A) ÷ (HACDE-G)→V;

R2=":V.
Goto 2;
Lbl 3(GB+1A) ÷ (HABDE)→W
"R3=":W
Goto 3;
Lbl 4: (GB+1A) ÷ (HABDE) →Y
"C1=":Y.
Goto 4:
Lbl 5:(J) ÷ (HABCD)→Z
"C2=":Z.
Goto 5;

The op-amp were guided to ensure stable D. C. biasing round all its terminal hence high pass at about (5Hz) were placed around the op-amp to guide tract its dc bias while that along the supplies were to maintain stability at high frequency and to filter distortion frequency in the supply lines. Speech fundamentals occur over a fairly limited range between about 125Hz and 250Hz. Vowels essentially contain the maximum energy and power of the voice, occurring over the range of 350Hz to 2000Hz. Consonant occurring over range of 1500Hz to 4000Hz contain little energy but are essential to intelligibility. The frequency range from 63 to 500Hz carries 60% of the power of the voice and yet contributes only 5% to the intelligibility. The 500Hz to 1KHz region produces 35% of the intelligibility, when then from 1K to 8KHz produces just 5% of the power of 60% of the intelligibility, Table 3.1 shows various vocals and their effect on sound. This helps in mixer design.

Filter Design Techniques (James, 1997) Oblique:
1). The band filter section were 0dB at fc
2). The design Q required for the desired bandwidths require a Q of 4.3185) without approximations, ISO approximations, that is for 793.7 it is not 800Hz.

Topology selection involved many trade-offs, but for special reasons MFB, type was preferred. The task was to design a second order unity gain filter with various corner frequencies. In addition, minimum phase behavior is an important criterion.

***Table 3.1** Vocal frequencies effect*

Frequencies Ranges (Hz)	Sound Effect and Importance	Effect
31 to 50	These frequencies give music a sense of power. If over emphasized they can make things middy and dull. Will also cloudy up some harmonic content.	
80 to 125	To much in this area produces excessive "boom"	Sense of power in some outstanding bass singers
160 to 250	This is the problem area of many mixers. Too much of this area can take away from the power of mix but is still needed for warmth. 160Hz is a pet peeve frequency. Also fundamentals of bass guitar and other bass instrument sit here	Voice fundamental
300 to 500	Fundamentals of string and percussion instruments	Important to voice quality (300 to 500)
400 to 1k	Fundamentals and harmonic of stings, keyboard and percussion. This is probably the most important area when to control or shape to a natural sound. The voice of an instrument is in the mids. Too much in the area can make instrument is in the mids. Too much in this area can make instruments sound horn-like.	Important for a natural sound. (630 to 1K). Too much boost in the 315 to 1K ranges produces a honky, telephones-like quality.
800 to 4K	This is a good range to accentuate instruments or warm them up. Too much in this area can make produce listening fatigue. Boosts in the 1K to 2K range can make instrument sound tinny.	1.25 to 4K and 5K to 8K are accentuation of vocal
4K to 10k	Accentuation of percussion, cymbals and snare drum. Playing with a 5k makes the oval sound more distance or transparent	1.25 to 4K and 5K to 8K are accentuation of vocal
8k to 20k	This area is often what defines the quality of receiving or mix. This area can also help define depth and air to mix. Too much can take from the natural sense of mix by becoming shrill and brittle.	1.25 to 4K and 5K to 8K are accentuation of vocal

Source :(11, 13)

3.2 Op-Amp Selections and Blue Print

The most important opamp parameter for proper filter functionality is the unity-gain bandwith. In general, the open loop gain should be 100 times (40dB above) the peak gain (Q) of a filter section to allow a maximum gain error of 1%

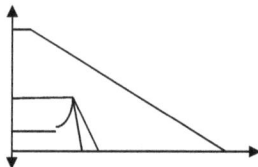

The following equations are good rules of thumb to determine the necessary unity gain bandwidth of an ap amp for an individual filter section
1. First order filter
   $Ft=100 Gain.fc$
2. Second order filter (Q<1):
   $Ft=100.Gain.fc.k_1$ with $K_1=\dfrac{fa}{fc}$
3. Second order filter (Q>1):
   $Ft=100.Gain. \sqrt{\dfrac{Q1-0.5}{Q1-0.25}}$

Besides good de performance, low noise, low signal distortion, another important parameter that determines the speed of an op amp is the slew rate (SR). For adequate ful power response, the slew rate would be greater than:

$SR=\pi.V_{pp}.f_c$

Another design rule was to ensure that the amplitude function be entirely from the band-pass filter function. Proper selection of the band pass filter topology involves many tradeoffs, not the least of which is cost. Several important things regarding the optimization combining characteristics had been incorporated:

1). Design the gain of the band pass filter sections to be precisely 0dB at fo
2). Design of the exact Q required for the desired bandwidth (for example, one third octave band width require a Q of 4.3185).
3). Design filter center frequencies should be exactly, and the skirts should cross at exactly dB for optimum combining.

Slider design noise problem was solved by taken resistor drawn from the summing nodes to ground equivalent valve of the parallel sliders; let R s represent the total resistance of the slider[13.] Since the center is grounded, R3/ 2 represent the amount resistance from each summing node to ground. Valve divided by n, where n equals the number of slides, the equivalent value. For 23 look sliders, this value equals approximately 2.2k ohm. The noise gain of each summing stage is now feedback resistor R divided by the parallel combination the input resistor R and the equivalent slider resistance since noise in a resistor is also proportional to the voltage across the resistor.

3.3   Overall Designed Circuit Diagram

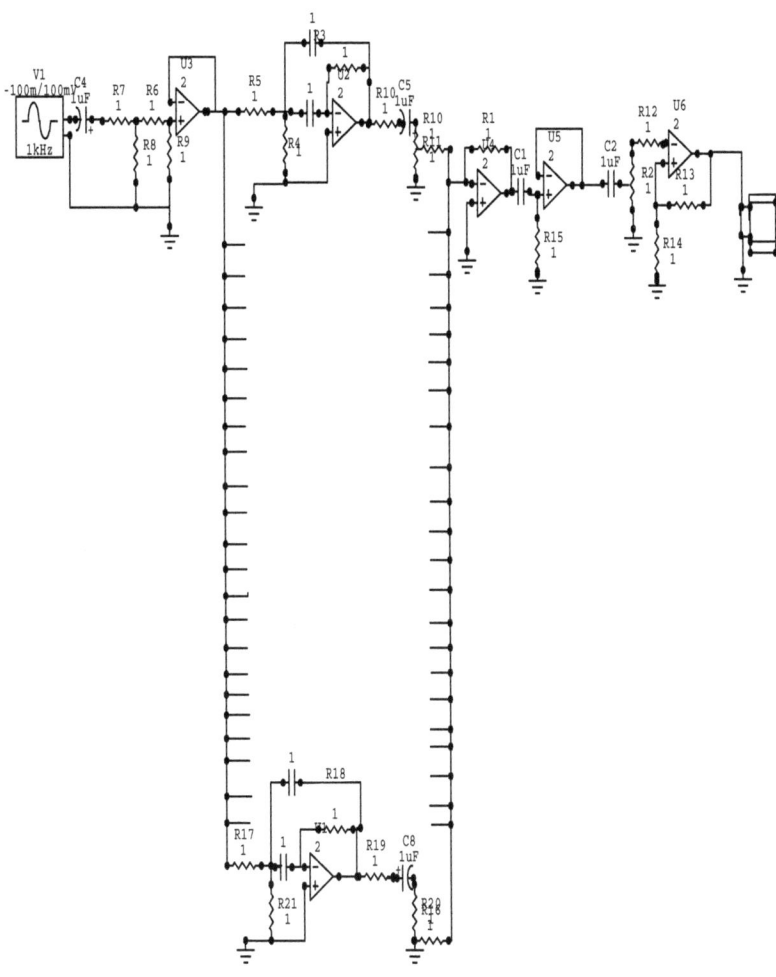

Overall designed circuit diagram

3.4 Cross over Design

The forth order butter-worth which is the fourth order Linkwitz-Riley [W] was selected for low (400Hz) and high crossover (5K or 4K) cuts while series would be used for the mid-range frequencies between 400Hz to 5Hz or 4Hz. Well, it is easier to make a loudspeaker capable of only handling a limited portion of the audio spectrum. The feedback provides continuous comparison between the input signal and what the amplifier is putting out. As this comparison is made errors between the real signal and any lack of faithfulness of the amplifier output tend be corrected through the feedback. The feedback allows cancellation of any change in the present gain. The concept gain bandwidth as specified in the database was explored to picture the op amp stability at about 30 KHz that implies the stable gain $\frac{100000}{30000}$ = 33.333 at a gain bandwidth of 1MHz

CHAPTER FOUR

4. ANALYSIS AND DISCUSSION

Computer simulation was done (see chapter five) on this network with mixed domino effects. The resultant amplitude was less but the ripple was less; the improvement was significantly better than for the single boost. Ground (at Vo) was used instead of cut summing amplifier.

4.1 Signal Coupling and Pre-Amplifier

The signal source input was first coupled via (7.12μf), ac-coupled the filter blocking any dc level in the signal while the op-amp (James, 1997) operates as a voltage follower and as an impedance converter (as Fig.21); the parallel circuit of the resistor, R_b (4.6K ohm, within attenuator), together with C_{in} create a high pass filter. These was that to avoid any effect on the low-pass characteristics of the in-coming waveform' the corner frequency of the input high-pass was low 1.58 to 5Hz) against the corner frequency of the actual low-pass; using fundamental resonance equation 4.1. The use of an input buffer causes no loading effects on the filter.

$$f_c = \frac{1}{2\pi fc} \quad \dots\dots\dots\dots\dots\dots\dots\dots\dots\dots\dots\dots\dots\dots\dots\dots\dots\dots\dots\dots\dots\dots\dots 4.1$$

$f_c \longrightarrow$ 3dB

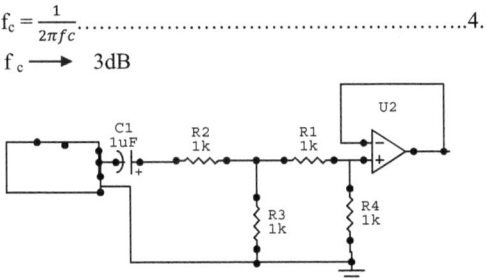

Figure 21: *Input circuit with attenuator*

In the case of a higher-order filter, all following filter stages receive their bias level from the preceding filter amplifier.

4.2 Multiple Feedback (MFB) Band-Pass Circuit

This was the heart of the whole, which involved rigorous calculations and simulation analysis to authenticate the design accuracies; however, biasing techniques of a Sallen-key and an MFB were shown in Fig.22, for single supply source, for single supply source.

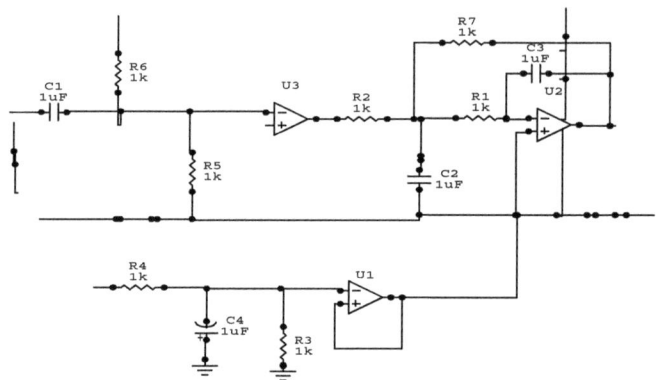

Figure 22. *Biasing a second order MFB filter biasing*

The input buffer decoupled the filter from the signal source. The filter itself was biased via the non-inverting amplifier input. For that purpose, the bias voltage was taken from the output of the generator with low output impedance. The op-amp operated as a difference amplifier and subtracts the bias voltage of the input buffer from the bias voltage of the generator, thus yielding a dc potential of zero input signal. The input capacitors of high-pass filter already provide the ac-coupling between filter and signal source. Both circuits use the generator biasing while the MFB, circuit is biased at the non-inverting amplifier input, while Sallen-key high-pass is biased via the only dc path available, while is R1. In the ac circuit, the input signals travel via the low output impedance of the op-amp to ground.

4.3   The Band Pass Design Calculated Values

Exemplified in Table 4.1 were the resulted computations from the filter program in chapter three.

Table 4.1: Bandpass filter values

S/N	Corner Lower Frequencies f1 (Hz)	Center Frequencies fc (Hz)	Corner Upper Frequencies f2(Hz)	R1 (Kn)	R2 (Kn)	R3 (Kn)	C1 (nanoF)	C2 (nanoF)	Quality Factor Q
1									
2		20		15.0	9.5	32.0	470.0	470.0	4.0
3	17.5	25	22.5	15.0	6.7	32.0	470.0	470.0	4.5
4	22.5	31	28.0	15.0	3.9	32.0	470.0	470.0	5.3
5	28.0	40	33.9	38.7	1.8	84.0	320.0	320.0	4.2
6	33.9	50	43.0	10.0	2.0	22.0	470.0	470.0	4.3
7	43.0	63	54.2	32.5	19.4	68.7	14.8	320.0	4.2
8	54.2	80	68.8	38.7	1.8	68.7	320.0	100.0	4.3
9	68.8	100	86.7	32.8	7.3	68.3	15.6	320.0	4.4
10	86.7	125	108.8	32.8	4.3	68.3	17.0	320.0	4.1
11	108.8	160	138.8	32.8	7.3	68.3	5.9	320.0	4.3
12	138.8	200	175.0	32.8	4.6	68.3	5.9	320.0	4.4
13	175.0	250	219.6	32.8	1.5	68.3	16.0	230.0	4.3
14	219.6	315	277.5	47.1	11.2	82.0	16.0	16.0	4.2
15	277.5	400	351.8	32.7	8.2	68.4	15.5	15.5	4.3
16	351.8	500	443.7	1.8	2.7	3.9	100.0	100.0	4.3
17	443.7	630	559.6	1.8	1.8	3.6	100.0	100.0	4.2
18	559.6	800	709.7	2.2	1.0	3.9	100.0	100.0	4.3
19	709.7	1000	895.0	15.0	7.38	33.3	9.7	9.7	1.7
20	895.0	2000	1598.7	47.2	1.5	101.0	5.6	5.6	1.3
21	1598.7	4000	3389.2	10.1	1.9	20.0	5.6	5.6	1.3
22	3389.2	8000	7185.4	3.3	1.0	6.2	5.6	9.1	1.3
23	7185.4	16000	15233.1	1.2	1.2	2.7	5.6	5.6	3.3
	15233.1	20000	20601.3	1.0	1.0	2.1	5.6	5.6	4.1
	20601.3		26276.0						

The next stage beyond the band pass bank is the crossover channels that the output response into speaker: bass/woofer, mid-range and tremble (tweeter). Low frequencies would damage tweeter; the butter-worthy fourth, Fig. 23 order used in this design to produce -24dB per octave would reduce lower frequencies- in high-pass- at a steeper rate than other lower orders.

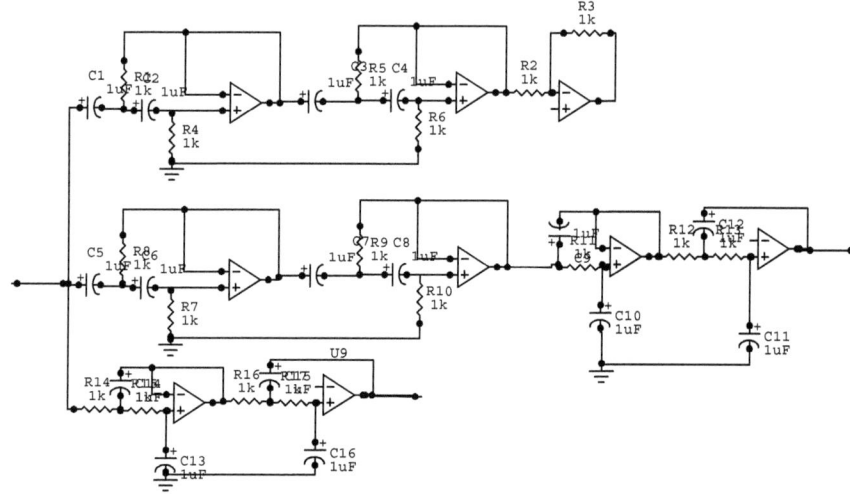

Figure 23. Crossover circuits

4.4 Output Section and Components Selection

The output load driver was such that could drive 4 ohms, 8ohms or 16 ohms speaker. Directly, the load was placed after the power amplifier (Fig. 24) couple to this was crossover with one power amplifier (Bohn, (1976) per channel was not constructed due to cost though designed. LM3876 does this well. The amplifier has the following operational parameters:

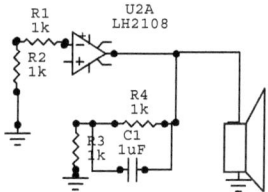

Figure 24. Power Amplifier circuit

Vcc/Vee (+/-)= 30.0volts
Abs.Max Voltage= 42.0 Volts
Voltage Headroom= 12.0 Volts
R_L (load impedance)= 8.00 ohms
R_f(feedback)= 100.0 Kilo ohm
R_1= 2.1 kilo ohm

C_1 = 470 uf
R_B = 1.0 kilo ohm
*R_{in} = 200 kilo ohm
*C_{in} = 1.00 uf
Gain = 48.62 v/v
Lower -3dB cutoff = 8.00 Hz
Input impedance = 20.0 kilo ohms
Power dissipation and thermal design are:
Maximum P_D = 22.8o watts
Total P_D/IC = 22.80 watts
$T_{ambient}$ = 25.0 °C

4.4.1 Capacitor Selection

The tolerance of the selected capacitors and resistors depends on the filter sensitivity and on the filter performance. Sensitivity is measure of the vulnerability of a filter's performance to changes in component values. The important filter parameters to consider are the corner frequency, fc and Q. when changes by 2% due to a 5% changes in the capacitance value, than the sensitivity of Q to capacitor changes are expressed as

$$S^Q_C \frac{2\%}{5\%} = 0.4 \frac{\%}{\%}$$

The following sensitivity approximations apply[6] to second-order Sallen-key and MFB filters:

$$S^Q_C = S^Q_R = S^{fc}_C = S^{fc}_R = \pm 0.5 \frac{\%}{\%}$$

Although 0.5%/% is small difference from ideal parameter, in the case of higher order filters, the combination of small Q and fc differences in each partial filter can significantly modify the overall filter response from its intended characteristics. Figure 4.4 and 4.5 show how an intended eighth-order Butterworth low-pass can into a low-pass with Tschebyscheff characteristic mainly due to capacitance changes from the partial filters. Capacitor selection is very important for a high-performance filter. Capacitor behavior can very significantly from ideal, introducing series resistance and inductance, which limit Q. in addition, nonlinearity of capacitance versus voltage causes distortion. Common ceramic capacitors with high dielectric constants, such as "high-k" types, can cause errors in filter circuits. The various temperature characteristics of ceramics capacitors are identified by a three-symbol code such as: COG, X7R, Z5U, and Y5V,COG-type ceramic capacitors are the most precise. Their nominal values range from 0.5 pf to approximately 47nf with initial tolerance from 0.25pf for smaller values and up to 1% for higher values. Their capacitance drift over temperature is typically 30ppm/0C. X7R-type ceramic capacitors range from 100pf to 2.2nF with an initial tolerance of +1% and a capacitance drift over temperature of 15%. For higher values, tantalum electrolytic capacitors should be used. Other precision capacitors are silver mica, metalized polycarbonate, and for high temperatures, polypropylene or polystyrene. Since capacitor values are not as finely subdivided as resistor values, the capacitor values need be defined prior to selecting resistors. If precision capacitors are not available to provide an accurate filter response, n then it is necessary to measure the individual capacitor values, and to calculate the resistor accordingly.

4.4.2 Component Valves

Resistor values should stay within the range of 1K ohms. The lower limit avoids excessive current draw from the op amp output, which is particularly important for single-supply op-amp in power-sensitive applications. Those amplifiers have typical output currents of between 1mA and 5mA. At supply voltages of 5V, this current translates to maintain of 1K ohms. The upper limit of 100K ohms was to avoid excessive resistor noise. Capacitor values can range from 1nF. The lower limit avoids coming too close to parasitic capacitance. If the common-mode input capacitance of the op-amp used in Sallen-key filter section[7], is close to 0.25% of C1, (C1/400), it must be considered for accurate filter response. The MFB topology, in comparison, does not require input capacitance compensation.

4.5   Peak Limiter/ Attenuator

The peak limiter works in such a way that when the voltage on the input exceeds the drop across the zener diode and diode it locks the maximum feedback voltage at that value causing the gain to decrease as the input voltage increases. This introduces odd harmonic distortion but that is unavoidable with all peak limiters. The distortion only occurs when the circuit is limiting the voltage since when the voltage is less than the

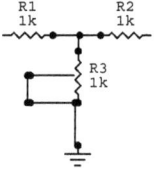
Figure 25. Signal noise attenuator

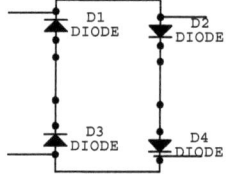
Figure 26. Clipping confiscator

Zener that path is open circuit. Resistance needs to be kept small to limit the distortion are kept small. Simple attenuator was implemented here, Fig. 25.

Volume Control

The volume and balance control is just a simple circuit. 100K ohms were there so that when the balance control was used to ensure that the sound level of the output remains relatively constant when the balance is adjusted without affecting the audio frequency range. 100k ohm also conveniently makes the adjustment more slow when close to the center of balance control. The volume control is just a simply voltage divide and balance of that an op-amp is used to give a large impedance so that it does not load the voltage divider. The resistors around the volume control cause a linear pot to appear approximately logarithmic which is good since general linear pots are less noisy then logarithmic pots.

Power Amplifier and Stability

Output stage is the portion which actually converts the weak input signal into a much more powerful "replica" which is capable of driven high power to speaker. This portion of amplifier typically uses a number of "power transistor" (or MOSFET's) and also responsible for generating the most heat in the unit (unless the amplifier happens to have a very bad power supply design). The power amplifier is a basic class A/B design. We want with a class A/B amp because of its low distortion and high efficiency compared to a class A or class B amplifier. The differences[12] between a class B amp and a class A/B is the current source that is realized by Q402, D401 and D402. This similar to class A design and is used ensure that Q403 and Q404 DO NOT completely turn off to reduce the distortion caused by the turn-on transients in the transistors. The constant current source ensures that Q401 is always on so that it acts like a diode which keeps Q403 biased properly at all times. Auto-oscillation might result if the power supply was not appropriately filtered yet a non-polarized capacitor valued about 100000Pf while the feedback 100 kilo ohm resistor necessitated about 310nf-530nFceramic cap (Fig. 25) to higher limit higher frequencies hence maintained gain stability.

Combining performance

The mixer is a convectional "virtual earth" type that minimizes interaction between the slid pots. The distortion stage incorporated diodes 1N4148 types as a clipping, Fig. 26, circuit. It makes sure that the voltage retains or allows be seeing or transmitting to the next stages is not > (0.7×2) since the signal is AC. It needed to be passed bi-directionally hence the Fig. 26

CHAPTER FIVE

5. Results and Conclusion

The phenomenon here was designed to flaunt the corporal depiction of each section of the project- simulated waveform of all the analysis so far and there technical Table specification.

5.1 Filter Shapes, Responses and Combining Action

The nature, shape and way in which individual tone filters combined, had a profound effect on the control provided by the refiner and on the resulting quality of sound. The majority of applications within the sound reinforcement, broadcast and recording fields, require a smooth and continuous equalization response curve in order to correctly contour the overall response characteristics of a sound system, loudspeaker, recording effect or audio channel. Each "cut" in the audio range should meet the designed quality it meant for and due to the "unfaithfulness" of components which might endure the bands optimum performance and selectivity; circuit simulation-shown in Fig. 27 to 47 was chief design tools to ensure conventionally to design target(s). To achieve this, the individual filters must be capable of combining smoothly together to result in a continuous response curve, free from shape discontinuities in order to avoid unwanted audible peaks or anomalies in the final sound attributes.

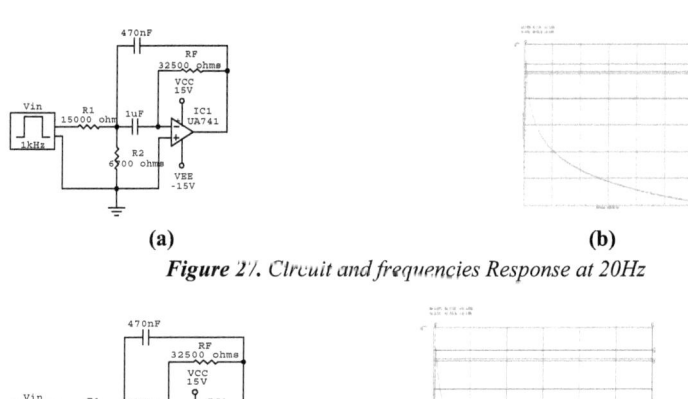

(a) (b)

Figure 27. Circuit and frequencies Response at 20Hz

(a) (b)

Figure 28. Circuit and frequencies Response at 25Hz

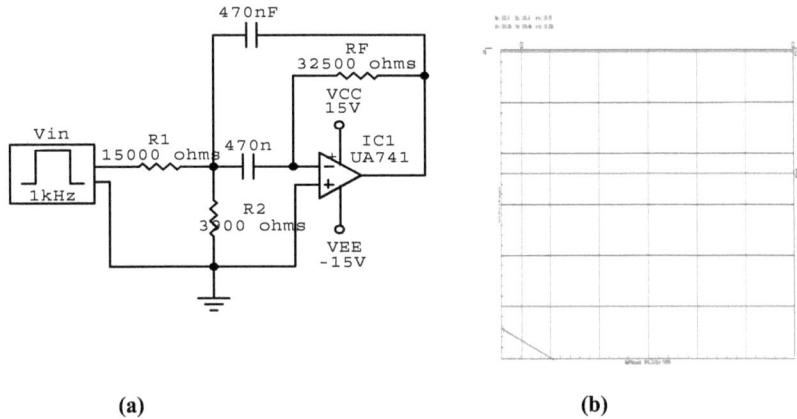

Figure 29. Circuit and frequencies Response at 31Hz, fc

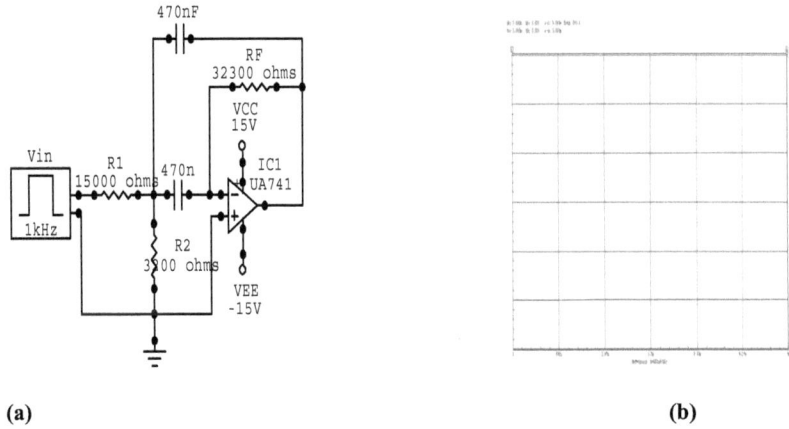

Figure 30. Circuit and frequencies Response at 40Hz, fc

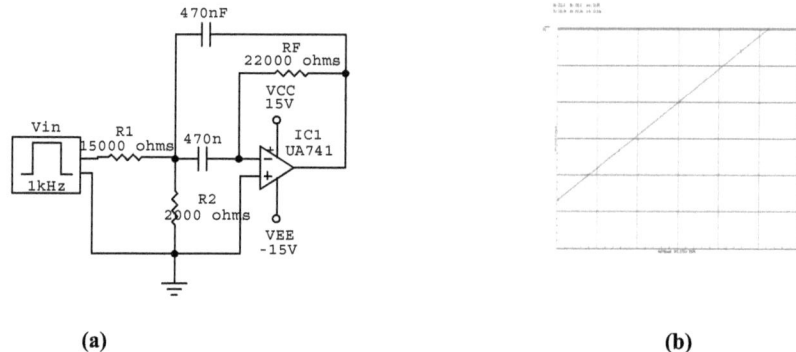

Figure 31. Circuit and frequencies Response at 50Hz, fc

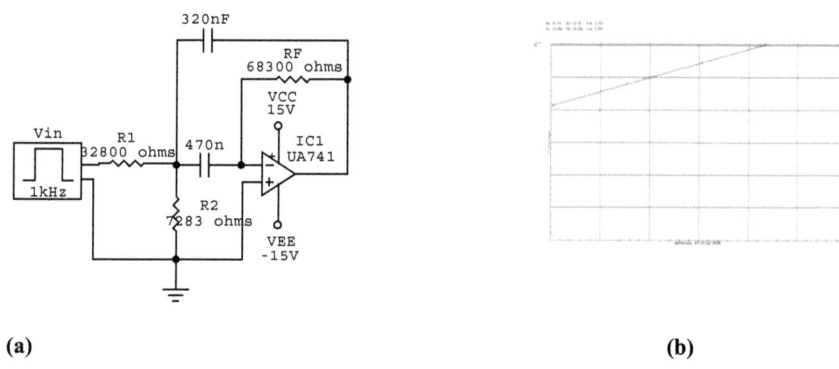

Figure 32. Circuit and frequencies Response at 80Hz, fc

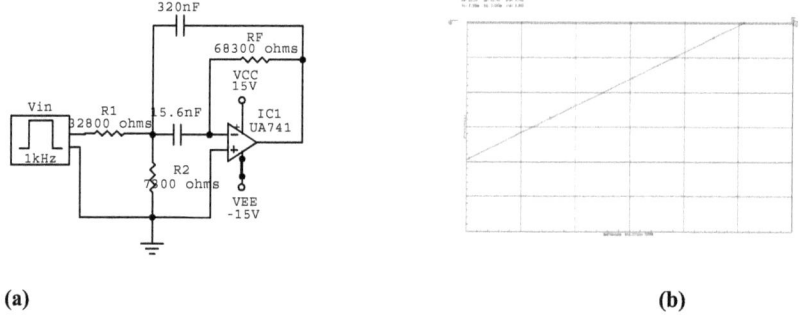

Figure 33. Circuit and frequencies Response at 100Hz, fc

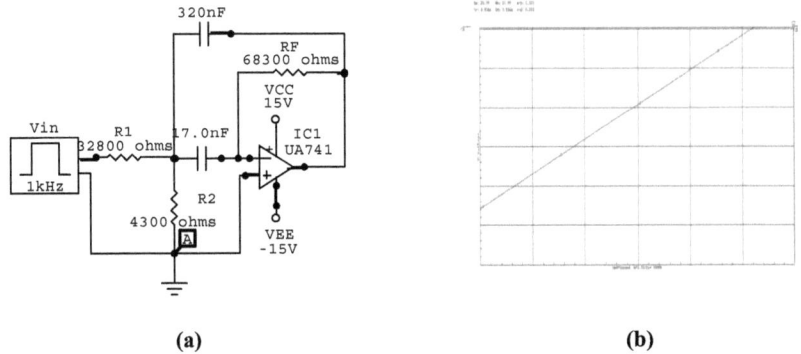

(a)   (b)

Figure 34. Frequencies Response at 125Hz, fc

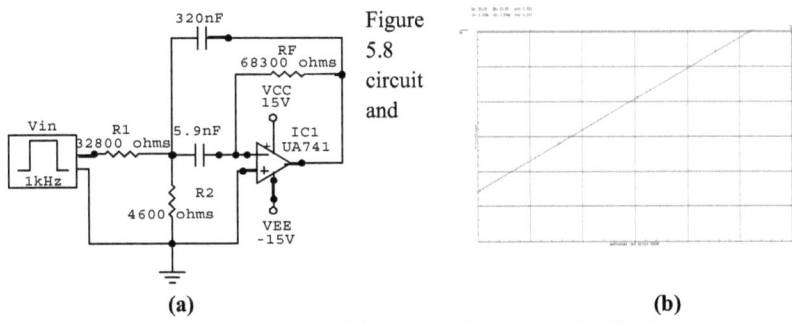

(a)   (b)

Figure 35. Circuit and frequencies Response at 200Hz, fc

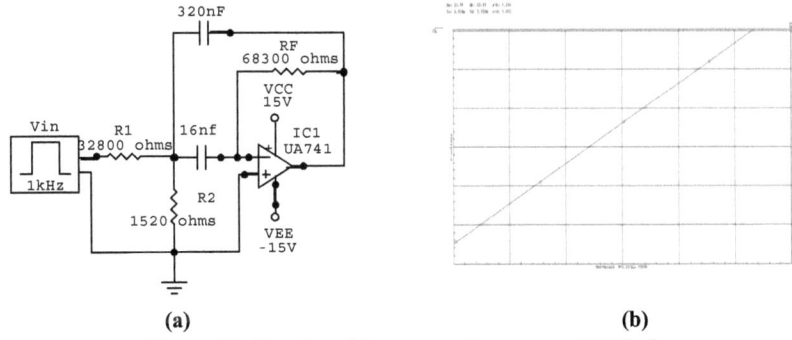

(a)   (b)

Figure 36. Circuit and frequencies Response at 250Hz, fc

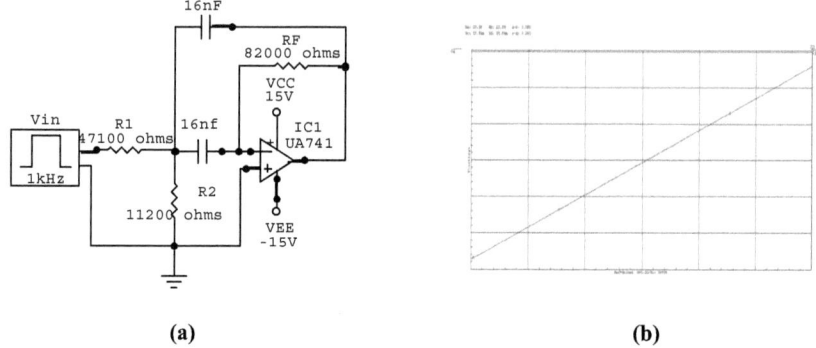

Figure 37. Circuit and frequencies Response at 315Hz, fc

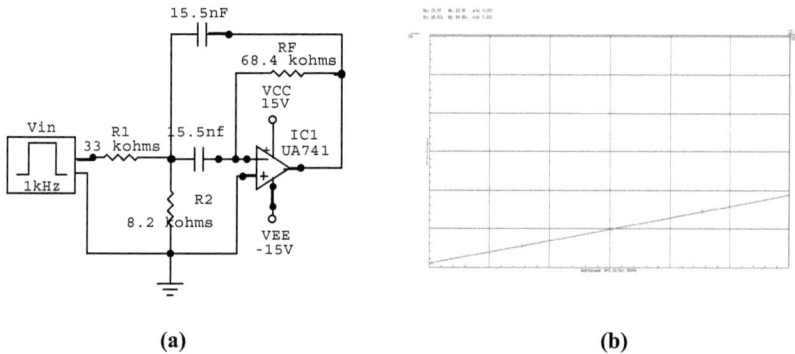

Figure 38. Circuit and frequencies Response at 400Hz, fc

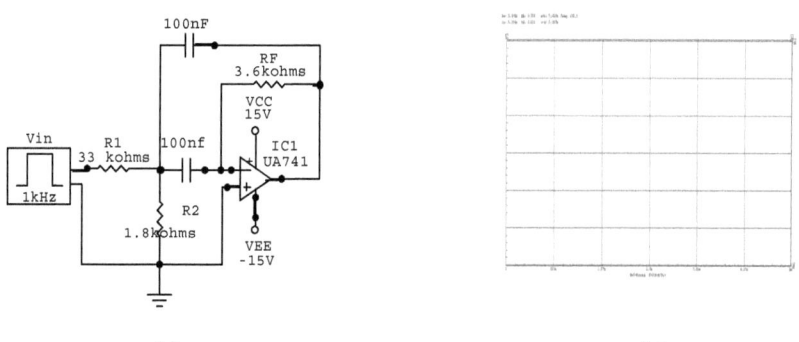

Figure 39. Circuit and frequencies Response at 630Hz, fc

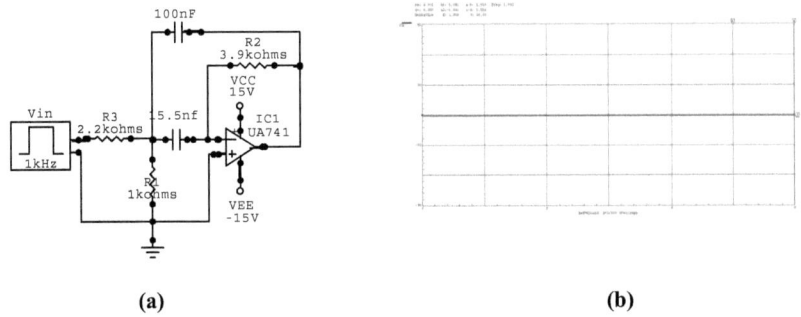

Figure 40. Circuit and frequencies Response at 800Hz, fc

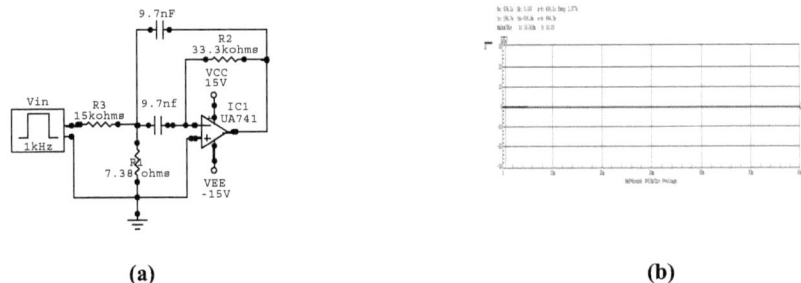

Figure 41. Circuit and frequencies Response at 800Hz, fc

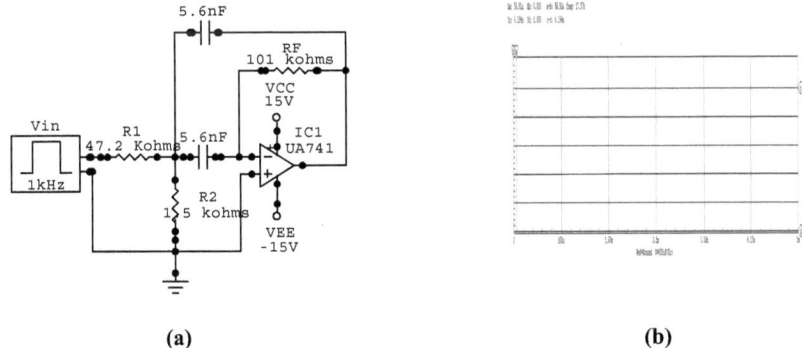

Figure 42. Circuit and frequencies Response at 2KHz, fc

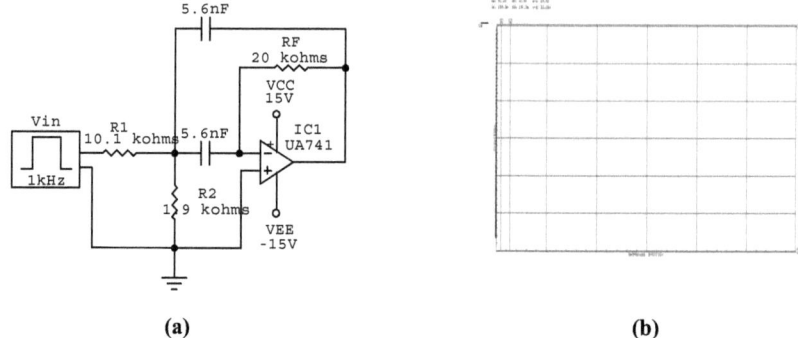

Figure 43. *Circuit and frequencies Response at 4KHz, fc*

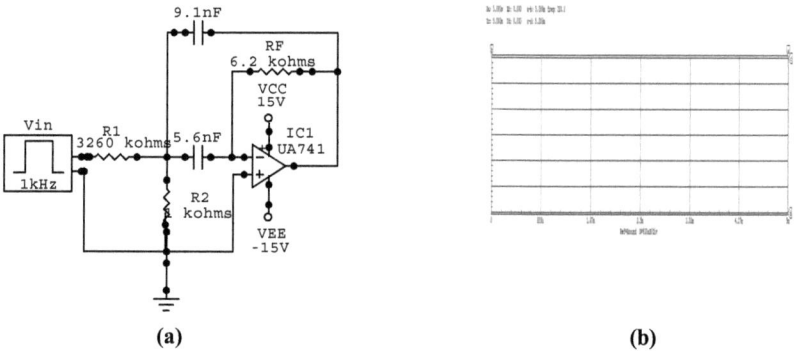

Figure 44. *Circuit and frequencies Response at 8KHz, fc*

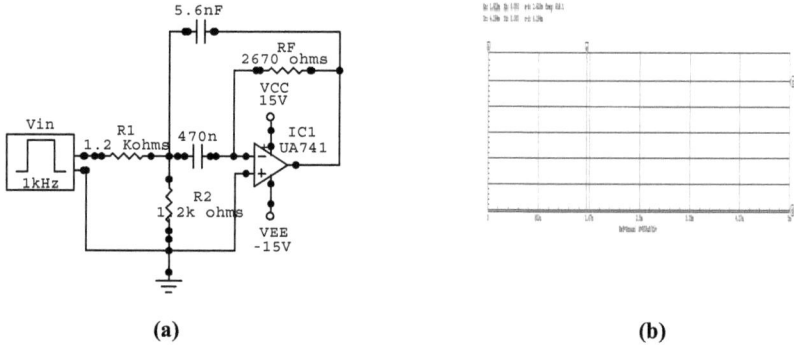

Figure 45. *Circuit and frequencies Response at 16 KHz, fc*

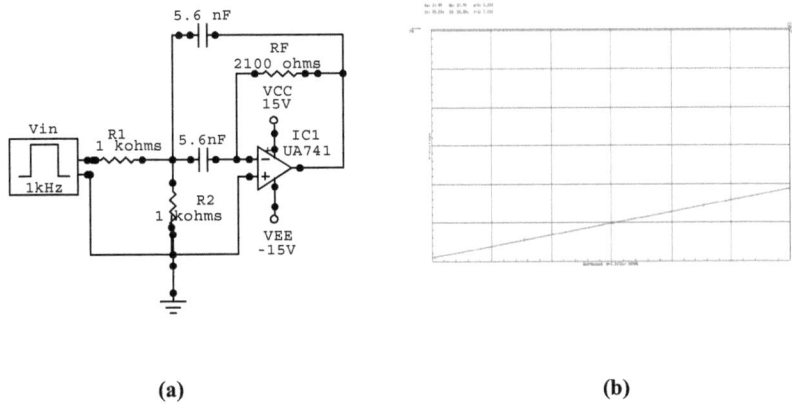

(a)   (b)

Figure 46. Circuit and frequencies Response at 20KHz, fc

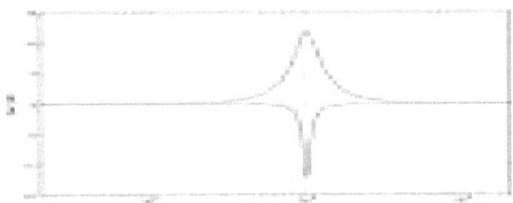

Figure 47. Single filter response curve (1/3 octave)

5.2 Noise Analysis

Noise can move from a power source (60Hz-hum); also from feedback networks, from mechanical system connected to electrical system, from stay capacitive, inductive efforts or possibly from a local signal source that is not properly shielded-the list is endless. 100uf capacitors were used to bypass the non-inverting input to earth for A.C. and help to reduce noise. The op-amp can be any common device for low frequencies, but at high frequencies (>2 kHz) a high- speed unit is required for best performance. Using distortion control will definitely increase the noise. The higher the resistance, the greater the noise (the laws of physics!) becomes. As described above reducing the value of the mixing resistor can heighten the frequency strength. All effort is made to confine each filter to its bandwidth to avoid distortion (clipping) if the input level is too high. So maximum input voltage is kept at 0.5 VRMS lower than this will provide a better safely margin and will ensure that clipping does not occur regardless of slider settings. Special consideration was ensured however to avoid any effect of the adjacent frequencies

Loading (at any stage) was eliminated as such as possible while audio output was effectively coupled or matched-'with minimum of × 10 rule'- to the next stage(s); stray capacitances on the board effects were similarly fudged.

The Audio Multitone Refiner Technical Specifications:

Inputs:

    Input impedance: 100KΩ
    Maximum input level: 1.5V
    Supply power voltage +/-15V
    (For voltage gain op-amps)

Outputs:

    Output impedance: 8Ω
    Maximum output level: 15.83watts into 8n or greater (in 4n)

General performance

    Frequency response: 17.5Hz-26Hz

Filters:

    Audio Multitone refiner: at quality factor between 5 and 1.5;
    Center frequency tolerance <5%
    Max boost/cut: switchable 6 or 12dB
    High pass filter: 1.58 Hz-frequency ranges

Power requirement

    Voltage: 180V-250V'50 Hz
    Consumption: about 30V
    Temperature:30 °C (with fanning).

Conclusion and recommendation

Apart from adding good crossover for the bass, mid-range and tremble, an overload LED per channel, which warns of impending overload at any point in the refiner; reliability control is then ensured

A signal-ground lift switch and an optional security cover to prevent unauthorized personnel from tampering with the control settings. This product can be improved upon to make local multitone refiner robust and stylish- to the same high electrical and mechanical standards as in all parametric and constant Q-equalizers.

Only high quality components are recommended without approximations. The actual measured capacitor values (not nominal values) should be directly used; in this way, an accurate filter response can be achieved with relatively inexpensive components. Some improvements that would make the amplifier to control that is available on the LM3876 were insertion of transistor switches. Also it would be nice if there were some switches that could remove the control and refiner if the listener does not want to affect the sound in anyway.

Deterrent measure should be taken on the semi-conductors while overheating the chip must be avoided. Heat sink (a good heat sink) is needed; a thin coat of flat black enamel paint seems to be effective. 100uf capacitors are highly recommended to bypass the non-inverting input to earth for A.C. and help to reduce noise.

Recommended capacitor types are: NPO ceramic, silver mica, metalized polycarbonate and, for temperatures up to 80 °C, polypropylene. To minimize the variations of fc and Q (COG) ceramic capacitors are recommended for high-performance filters. These capacitors hold their nominal value over a wide temperature and voltage range.

For high performance filter, 0.1% resistors are recommended. Approximate or round off numbers should be discarded as possible. The skirts must cross at exactly corner frequencies. Usage of two series summing networks (three or more do not contribute any better result. Future pollster should look for way of adding level meter display: headroom indication and clip led light for approximately 500ms will add longevity and efficiency to the device.

Convincingly, as it is obviously seen, this project suffered pecuniary patronage, and time hampering played a requisite role; well, this is a stupendous challenge to student(s) of electronic to see this research work very desirable-in the indigenous technology development in Nigeria. Much challenges, to reduce noise due to variable resistors and to develop a better mixing console for audio signal with 120dB or more signal to noise ratio gain; so fascinating indeed but not with ease.

REFERENCES AND BIBLIOGRAPHY

1) Bohn D.A (1976), Audio handbook (National Semi-Conductor Corp.,) pp. 2-53-2-5
2) Franko, S.(1988) "Design With Operational Amplifiers And Analog Integrated Circuits", Mcgraw- Hill.
3) James, Garratt (1997) "Design and Technology" Low Price Edition, Cambridge.
4) James W.W. and Susan,A.R. (1999) "Electronic Circuit" Sixth Edition NJ, Prentice Hall.
5) Tab Books INC. (1982) "The Giant Book Of Blue Ridge Summit, Tab Book Inc.
6) http://besi.de
7) http://www.boutell.com/gd
8) http://www.carstero.com
9) http://www.electronickits.com/kit
10) http://www.geofex.com
11) http://www.max-ic.com
12) http://www.prenhall.com/boylestad
13) http://www.rfcafe.com/references/electrical/cheby poles.htm
14) http://www.swrthmore.edu
15) http://www.thole.org
16) http://www.nuhertz.com/software/software-modules/active-filter-module
17) http://www.filter-solutions.com/active.html